Emergence, Complexity and Computation

Volume 48

The Emergence, Complexity and Computation (ECC) series publishes new developments, advancements and selected topics in the fields of complexity, computation and emergence. The series focuses on all aspects of reality-based computation approaches from an interdisciplinary point of view especially from applied sciences, biology, physics, or chemistry. It presents new ideas and interdisciplinary insight on the mutual intersection of subareas of computation, complexity and emergence and its impact and limits to any computing based on physical limits (thermodynamic and quantum limits, Bremermann's limit, Seth Lloyd limits…) as well as algorithmic limits (Gödel's proof and its impact on calculation, algorithmic complexity, the Chaitin's Omega number and Kolmogorov complexity, non-traditional calculations like Turing machine process and its consequences,…) and limitations arising in artificial intelligence. The topics are (but not limited to) membrane computing, DNA computing, immune computing, quantum computing, swarm computing, analogic computing, chaos computing and computing on the edge of chaos, computational aspects of dynamics of complex systems (systems with self-organization, multiagent systems, cellular automata, artificial life,…), emergence of complex systems and its computational aspects, and agent based computation. The main aim of this series is to discuss the above mentioned topics from an interdisciplinary point of view and present new ideas coming from mutual intersection of classical as well as modern methods of computation. Within the scope of the series are monographs, lecture notes, selected contributions from specialized conferences and workshops, special contribution from international experts.

Indexed by zbMATH.

Vincenzo Manca · Vincenzo Bonnici

Infogenomics

The Informational Analysis of Genomes

Springer

Vincenzo Manca
Department of Computer Science
University of Verona
Pisa, Italy

Vincenzo Bonnici
Department of Mathematics, Physics
and Computer Science
University of Parma
Parma, Italy

ISSN 2194-7287 ISSN 2194-7295 (electronic)
Emergence, Complexity and Computation
ISBN 978-3-031-44500-2 ISBN 978-3-031-44501-9 (eBook)
https://doi.org/10.1007/978-3-031-44501-9

This Springer imprint is published by the registered company Springer Nature Switzerland AG
The registered company address is: Gewerbestrasse 11, 6330 Cham, Switzerland

We dedicate this book to our nearest nodes in the tree of life to which we belong: our parents, our children, and grandchildren (of the oldest author).

Preface

This book originates in the courses that authors took in the last ten years at the universities of Verona and Parma, and before at the University of Udine and Pisa.

They have been on very different subjects centered on information, in many scientific fields: computation, formal representation, unconventional computing, molecular biology, artificial intelligence. The names of courses were: Information Theory, Languages and Automata, Unconventional Computing, Discrete biological Models, Programming Laboratory, Data Structures, Computational Biology. The book does not collect all the topics of these courses, but surely influenced by them, it is focused on genomes.

The conviction that we maturated during our teaching and related research, during last years, was the centrality of information as the common factor of all subjects of our interests. Information is the most recent scientific concept. It emerged in the last century, by revolutionizing the whole science. The reason of this is evident. Namely, all scientific theories develop by collecting data from observed phenomena, and these data support theories which explain phenomena and provide new scientific knowledge.

Information is inside data, but it does not coincide with them. It assumes and requires data, in order to be represented, but is independent from any specific data representation, because it can be always equivalently recovered by encoding data into other data.

This is the reason of its deep almost evanescent nature, but also of its fluidity, flexibility, adaptability, and universality. Information can always be linearized into strings. Genomes, which are the texts of life, are strings. But these linear forms of representation are managed, elaborated, transformed, and become meaningful, through networks, made of elements connected by relations, at different levels of complexity. Our brains are a particular case of such networks which develops through sensory, linguistic, social, and cultural information. In the dynamics of the communication, the ethereal essence of information flows and determines the various kinds of reality of our knowledge universe.

This book tells small fragments of this big story, that we collected and partially understood during our scientific adventure.

Pisa, Italy Vincenzo Manca
Parma, Italy Vincenzo Bonnici

Acknowledgments We express our gratitude to the students of the universities of Verona, Pisa, and Parma, which in the last ten years followed our courses in the curricula of Computer Science, Biological Sciences, Computational Biology, and Mathematics, at different levels. Their comments and suggestions improved the quality of the notes from which this book originates.We thank also our colleagues: Giuditta Franco, Giuseppe Scollo, Luca Marchetti, Alberto Castellini, Roberto Pagliarini, Rosario Lombardo, of the University of Verona, who collaborated with us in writing many papers in the research area of the book.

Contents

Acronyms

2LE	Binary Logarithmic Entropy
AC	Anti-entropic Component
AEP	Asymptotic Equipartition Property
AI	Artificial Intelligence
AMD	Average Multiplicity Distribution
ANN	Artificial Neural Network
BB	Bio Bit
CSO	Constraint Optimization Problem
EC	Entropic Component
ECP	Entropic Circular Principle
EED	Empirical Entropy Distribution
EP	Equipartition Property
ESA	Enhanced Suffix Arrays
ghl	Greatest Minimal Hapax Length
GPT	Generative Pre-trained Transformer
LCP	Longest Common Prefix
LE	Logarithmic Entropy
LG	Logarithmic Length
LX	Lexical Index
mcl	Maximal Complete Length
mfl	Minimum Forbidden Length
mhl	Minimum Hapax Length
mrl	Maximum Repeat Length
nhl	No Hapax Length
nrl	No Repeat Length
PSO	Particle Swarm Optimization
RDD	Recurrence Distance Distribution
WCMD	Word Co-Multiplicity Distribution
WLD	Word Length Distribution
WMD	Word Multiplicity Distribution

Chapter 1
The Infogenomics Perspective

Prologue

This book presents a conceptual and methodological basis for the mathematical and computational analysis of genomes, as developed in publications [1–13]. However, the presentation of the following chapters is, in many aspects, independent from their original motivation and could be of wider interest.

Genomes are containers of biological information, which direct the cell functions and the evolution of organisms. Combinatorial, probabilistic, and informational aspects are fundamental ingredients of any mathematical investigation of genomes aimed at providing mathematical principles for extracting the information that they contain.

Basic mathematical notions are assumed, which are recalled in the next chapter. Moreover, basic concepts of Chemistry, Biology, Genomics [14], Calculus and Probability theory [15–17] are useful for a deeper comprehension of several discussions and for a better feeling of the motivations that underlie many subjects considered in the book.

A key concept of this book is that of distribution, a very general notion occurring in many forms. The most elementary notion of distribution consists of a set and a "multiplicity" function that associates a number of occurrences to each element of the set. Probability density functions or distributions in the more general perspective of measure theory generalize this idea. Distributions as generalized functions are central in the theory of partial differential equations. The intuition common to any notion of distribution is that of a "quantity" spread over a "space", or parts, of a whole quantity, associated to locations. A probability distribution is a way of distributing possibilities to events. Probability and Information are for many aspects two faces of the same coin, and the notion of "Information Source", which is the starting point of Shannon's Information Theory, is essentially a probability distribution on a set of data.

The topics presented in the book include research themes developed by authors in the last ten years, and in many aspects, the book continues a preceding volume published by Vincenzo Manca, *Infobiotics: Information in biotic systems, Springer, 2013* [4]. We could shortly say that, abstractly and concisely, the book [4] investigates

V. Manca and V. Bonnici, *Infogenomics*, Emergence, Complexity and Computation 48,
https://doi.org/10.1007/978-3-031-44501-9_1

the relationship between rewriting and metabolism (metabolic P systems are the formal counterpart of metabolic systems), while the present book investigates the relationship between distributions and genomes.

The main inspiring idea of the present book is an informational perspective to Genomics. Information, is the most recent, among the fundamental mathematical and physical concepts developed in the last two centuries [18, 19]. It has revolutionized the whole science, and continues, in this direction, to dominate the trends of contemporary science. In fact, any discipline collects data from observations, by providing theories able to explain, predict, and dominate natural phenomena. But data are containers of information, whence information is essential in any scientific elaboration.

The informational viewpoint in science becomes even more evident for the sciences of life, which after molecular biology, have revealed as living organisms elaborate and transmit information, organized in molecular structures. Reproduction is the way by means of which information is transmitted along generations, but, in the same moment that biological information is transmitted, random variations alter a small percentage of the message passed to the reproduced organism.

Any reproductive system is *reflexive*, in the sense that it includes a proper part containing the information for reconstructing the whole organism. Thus, reflexivity is a purely informational concept, and DNA molecules are the basilar forms of reflexive structures, where reflexivity is realized by pairing two strands of biopolymers that can be recovered, by a template-driven copying process, from only one of them.

Genomes evolved from very short primordial biopolymers, proto-genomes, that assumed the function of biological memories. Enzymes, which direct, as catalysts, metabolic reactions, are the first molecular machines, but as any physical complex structure, they are subjected to destructive forces acting over time. The only way of keeping their specific abilities is the possibility of maintaining traces of them. Thus, proto-genomes probably originate as enzymatic memories, sequences including "copies" of the enzymes of proto-cells (membrane hosting substances and reactions among them). These memories, passing along generations, allowed life to develop and evolve, without losing the biological information cumulated along generations.

Genomes are "long" strings, or texts, of millions or billions of letters. These texts describe forms of life and even the history of life that produced them. Therefore, genomes are biologically meaningful texts. This means that these texts postulate languages. The main challenge for the biology of the next centuries is just discovering these languages. Specific steps in this direction are the following: (1) to identify the minimal units conveying information and their categories; (2) to discover the syntax expressing the correct mechanisms of aggregation of genomic units, at their different levels of articulation; (3) to understand the ways and the contexts activating statements that direct the (molecular) agents performing specific cellular tasks; (4) to determine the variations admitted in the class of individual genomes of a given species; (5) to reconstruct the relationships between genome structures and the corresponding languages that these structures realize in the communication processes inside cells.

Genomes are texts, but in a very specific sense, which is hardly comparable with the usual texts telling stories or describing phenomena. Maybe, a close comparison could be done with a legal code, prescribing rules of behaviour, with a manual of the use of a machine, or with an operative system of a computer. However, this special kind of manual/program is a sort of "living manual", cabled in the machine, as the interpreter of a programming language, but able to generate new machines. This fact is directly related to the cruciality of reproduction. Genomes generate statements assimilable to linguistic forms of usual languages, which instruct cell agents and, in some particular contexts, can change just parts of the same genomes directing the play, but somewhat paradoxically, they maintain a memory of their past history, but also of the mechanisms allowing their future transformations.

In fact, genomes structures and related languages provide an intrinsic instability promoting a small rate of change pushing toward new structures. This means that there are general genome built-in rules of internal variation and reorganization. Genealogies of life are firstly genealogies of genomes, very similar to those of natural languages. In this way, from a biological species new species derive, through passages called *speciations*. A new species arises when individual variations, in restricted classes of individuals, cumulate in such a way that some internal criteria of coherence require a new organization incompatible with the genome of the ancestor organism.

These dynamics require two kinds of interactions, one between species and individuals, and another between syntax and semantics of the genome languages. The genome of a species is an abstract notion, because, in the concrete biological reality, only individual genomes exist, and the genome of a species is the set of genomes of all individuals of that species. If some specific attributes have a range of variability, in the genome of the species these attributes become variables. Such a kind of abstract genome is a pattern of individual genomes. Fisher's theorem on natural selection (due to Ronald Fisher, 1890–1962) establishes that the time variation rate of genomes, of a given species, is proportional to the genome variability among the individuals of that species. In other words, the more genomes are variegate in the individuals living at some time, the more, on average, they change along generations.

Variations in individual genomes promote new interactions between syntax and semantics of genome languages, where semantics "follows" syntax. In other words, syntax admits some statements that even if well-formed can be meaningless. However, under the pressure of some circumstances, certain statements, casually and temporarily, assume new meanings that in cell interactions "work", becoming seeds of new semantic rules. If these rules are reinforced by other situations of success in cellular communication, they eventually stabilize. Speciation occurs when the number of genome variations overcomes a given threshold so that the language underlying the genomes results differently with respect to those of ancestor genomes. At that moment, the resulting organisms represent a new branch in the tree of life.

This "linguistic" scenario is only evocative of possible investigations, however, it is a good argument for explaining why considering genomes as "texts of life" is fascinating and could be one of the most promising approaches to the comprehension of the evolutionary mechanisms. In this regard, a clue about *life intelligence* resides in the relationship between chance and purpose. Randomness provides a powerful generator of possibilities, but the navigation in the space of possibilities selects trajectories that result in more convenience to realize some finalities. Probably, this form of intelligence, is spread evenly in the primordial genomes (as rules of syntactical coherence?). This possibility could explain the efficiency of evolution in finding the best solutions for life development. This theme is present since Wiener's pioneering research in Cybernetics ("Purpose, Finality and Teleology" is the title of Wiener's paper in 1943, with Arturo Rosenblueth and Julian Bigelow, anticipating his famous book published in 1948). Random genomes, which will be considered in the chapter "Information and Randomness" of this book, show a purely informational property of random texts.

This short introduction wants to emphasize the role that information plays in biological systems. The awareness of this fact persuaded us that a deep understanding of life's fundamental mechanisms has to be based on the informational analysis of biological processes. In this spirit, the following chapters outline general principles and applications of information theory to genomes.

Finally, a general remark on information concerns its relationship with knowledge. It is important to distinguish between the two notions. Knowledge is surely based on information, but a collection of data is something completely different from knowledge. Knowledge appears when data are interpreted within theories where they fit, by revealing an internal logic connecting them. But theories do not emerge automatically from data, conversely, data are explained when they are incorporated within theories, which remain scientific creations of human (or artificial?) intelligence.

References

1. Manca, V.: On the logic and geometry of bilinear forms. Fundamenta Informaticae **64**, 261–273 (2005)
2. Manca, V., Franco, G.: Computing by polymerase chain reaction. Math. Biosci. **211**, 282–298 (2008)
3. Castellini, A., Franco, G., Manca, V.: A dictionary based informational genome analysis. BMC Genomics **13**, 485 (2012)
4. Manca, V.: Infobiotics: Information in Biotic Systems. Springer, Berlin (2013)
5. Manca, V.: Infogenomics: Genomes as Information Sources. Chapter 21, pp. 317–324. Elsevier, Morgan Kauffman (2016)
6. Bonnici, V., Manca, V.: Infogenomics tools: a computational suite for informational analysis of genomes. J. Bioinform. Proteomics Rev. **1**, 8–14 (2015)
7. Bonnici, V., Manca, V.: Recurrence distance distributions in computational genomics. Am. J. Bioinform. Comput. Biol. **3**, 5–23 (2015)
8. Manca, V.: Information theory in genome analysis. In: Membrane Computing, LNCS, vol. 9504, pp. 3–18. Springer, Berlin (2016)
9. Bonnici, V.: Informational and Relational Analysis of Biological Data. Ph.D. Thesis, Dipartimento di Informatica Università di Verona (2016)
10. Bonnici, V., Manca, V.: Informational laws of genome structures. Sci. Rep. **6**, 28840 (2016). http://www.nature.com/articles/srep28840. Updated in February 2023
11. Manca, V.: The principles of informational genomics. Theor. Comput. Sci. (2017)

12. Manca, V., Scollo, G.: Explaining DNA structure. Theor. Comput. Sci. **894**, 152–171 (2021)
13. Bonnici, V., Franco, G., Manca, V.: Spectrality in genome informational analysis. Theor. Comput. Sci. (2020)
14. Lynch, M.: The Origin of Genome Architecture. Sinauer Associate. Inc., Publisher (2007)
15. Aczel, A.D.: Chance. Thunder's Mouth Press, New York (2004)
16. Feller, W.: An Introduction to Probability Theory and Its Applications. Wiley, New York (1968)
17. Schervish, M.J.: Theory of Statistics. Springer, New-York (1995)
18. Brillouin, L.: Scienze and Information Theory. Academic, New York (1956)
19. Brillouin, L.: The negentropy principle of information. J. Appl. Phys. **24**, 1152–1163 (1953)

Chapter 2
Basic Concepts

2.1 Sets and Relations

A set is a collection of elements considered as a whole entity characterized only and completely by the elements belonging to it. Braces are used for enclosing a finite list of elements of a set, or for enclosing a variable x and (after a bar $|$) a property $P(x)$ true when x assumes values that are elements of the set: $\{x \mid P(x)\}$.

Two basic relations $\in$ and $\subseteq$ denote when an element belongs to a set, and when a set is included in another set (all elements of the first set belong also to the second set). A special set is the empty set denoted by $\emptyset$ that does not contain any element.

On sets are defined the usual boolean operations of union $\cup$, intersection $\cap$ and difference–. The powerset $\mathbb{P}(A)$ is the set of all subsets of A.

The pair set $\{a, b\}$ has only a and b as elements.

The ordered pair (a, b) is a set with two elements, but where an order is identified. A classical way to express ordered pair is the following definition due to Kuratowski:

$$(a, b) = \{a, \{a, b\}\}$$

where the first element a is the element that is a member of the set and of the innermost set included in it, whereas the second element b belongs only to the innermost set.

The cartesian product $A \times B$ of two sets is the set of ordered pairs where the first element is in A and the second is in B.

The notion of ordered pair can be iteratively applied for defining n-uples for any n in the set $\mathbb{N}$ of natural numbers. In fact, an n-uple is defined in terms of pairs:

$$(a_1, a_2, a_3, \ldots, a_n) = (a_1, (a_2, (a_3, \ldots a_n))).$$

Logical symbols $\forall, \exists, \vee, \wedge, \neg, \rightarrow$ abbreviate *for all, there exists, or, and, not, if-then*, respectively.

A k-ary relation R over a set A is a set of k-uples of elements of A. $R(a_1, a_2, \ldots, a_k)$ is an abbreviation for $(a_1, a_2, \ldots, a_k) \in R$.

V. Manca and V. Bonnici, *Infogenomics*, Emergence, Complexity and Computation 48,
https://doi.org/10.1007/978-3-031-44501-9_2

Binary relations are of particular importance and among them equivalence and ordering relations, with all the concepts related to them. Often $R(a, b)$ is abbreviated by aRb.

A binary relation over a set A is an equivalence relation over A if the following conditions hold:

$aRa \quad \forall\, a \in A$ (Refexivity)
$aRb \Rightarrow bRa \quad \forall\, a, b \in A$ (Symmetry)
$aRb \wedge bRc \Rightarrow aRc\ \forall \quad a, b, c \in A$ (Transitivity)

Given an equivalence class over a set A and an element a of A the set $[a]_R = \{x \in A \mid aRx\}$ is the equivalence class of a. The set $\{[x]_R \mid x \in A\}$ is the quotient set of A with respect to R, denoted by $A/_R$. This set is a partition of A, that is, the union of all its sets coincides with A (it is a covering of A) and these sets are non-empty and disjoint (with no common element).

A binary relation over a set A is an ordering relation over A if the following conditions hold:

$aRa \quad \forall\, a \in A$ (Reflexivity)
$aRb \wedge bRa \Rightarrow a = b \quad \forall\, a, b \in A$ (Antisymmetry)
$aRb \wedge bRc \Rightarrow aRc \quad \forall\, a, b, c \in A$ (Transitivity)

An ordering relation is partial if two elements exist such that both aRb and bRa do not hold. In this case, a, b are incomparable with respect to R. An order is linear or total if for all two elements $a, b \in A$ at least one of the two cases aRb and bRa holds. Given a subset B of A, the element $m \in B$ is the minimum of B for R if mRb for all $b \in B$, while m is minimal in B if no $b \in B$ exist such that bRm. Analogously $M \in B$ is the maximum of B for R if bRM for all $b \in B$, and M is maximal in B if no $b \in B$ exists such that MRb.

Given a subset B of A the element $m \in A$ is a lower bound of B, with respect to R, if mRb for all $b \in B$. Analogously $M \in B$ is an upper bound of B with respect to R if bRM for all $b \in B$. In these cases, m is the greatest lower bound of B with respect to R if m is the maximum in the set of the lower bounds of B, and analogously M is the lowest upper bound of B, with respect to R, if M is the minimum in the set of the upper bounds of B.

A k-operation ω over a set A is a $(k+1)$-ary relation over A where the last argument depends completely on the others. In other words, if two $(k+1)$-uples of the relation coincide on the first k components, then they coincide on the last argument too. The last argument is also called the result of the operation ω on the first k arguments $a_1, a_2, \ldots, a_k$ and is denoted by $\omega(a_1, a_2, \ldots, a_k)$.

A function of domain A and codomain B

$$f : A \rightarrow B$$

(from A to B) is identified by the two sets A, B and by an operation f such that when f is applied to an element a of A, then the result $f(a)$ is an element of B, also called the image of a according to f and if $f(a) = b$, element a is also called the counter-image or inverse image of b. A function is injective when distinct elements of the domain have always distinct images in the codomain. A function is surjective when all the elements of B are images of elements of A, A function is 1-to-1 or bijective when it is injective and surjective. Given a subset $C \subseteq B$ the set of inverse images $\{x \in A \mid f(x) \in C\}$ is denoted by $f^{-1}(C)$.

Set-theoretic concepts and notation are the basic language of mathematics, developed in the last two centuries [1, 2]. Within the set theory, Cantor developed a rigorous mathematical analysis of infinity, by showing the rich articulation of this concept when it is mathematically formalized (transfinite, ordinal and cardinal numbers are sets). Set theory is also the basis of general topology, where general notions of space can be developed. Nowadays, almost all mathematical theories can be expressed in the set-theoretic language (see [3, 4] for introductory texts to set theory). For the needs of our discourse, the basilar concepts given here are enough. The foundational power of sets will not appear explicitly in the chapters of this book. However, it is important to realize that the expressiveness and universality of set-theoretic language is mainly due to the conceptual strength of the set theory, which in its advanced topics remains one of the most complex and fascinating mathematical theories.

A finite sequence of length n over a set A is a function from (positions) $\{1, 2, \ldots, n\}$ to A.

A finite multiset of size n over a set A can be considered as a function from A to natural numbers assigning a multiplicity to any element of the set A, such that the sum of all multiplicities is equal to n.

An infinite sequence, denoted by $(a_n \mid n \in \mathbb{N})$ has the set of natural numbers as a domain. The same notation with a generic set I of indexes denotes a family over a set A (indexed by I). An alphabet A is a finite set of elements, called symbols. Finite sequences of symbols are also called strings.

A variable is an element x which is associated with a set A, called the range of x, where the variable takes values in correspondence to some contexts of evaluation. The notion of variable is strictly related to that of function, but in a variable is stressed its range rather than the correspondence between contexts of evaluation (corresponding to the domain of a function) and the values assumed in the range (corresponding to the codomain of a function). In fact, for a variable, often, the mechanism determining the values is unknown or is not relevant.

If a variable x takes values in correspondence to the values of another variable, for example, a variable t of time, then the value assumed by x are denoted by $x(t)$, and the correspondence,

$$t \mapsto x(t)$$

determines a function from the range of t to the range of x.

When we apply operations to elements, in a composite way, for example:

$$\omega(x, \eta(y)))$$

where ω is a binary operation and η is a unary operation, then the expression above denotes a value in correspondence to the values of variables x, y. Parentheses are essential for denoting the order of applications of operations. In the case of the expression above the innermost parentheses are related to the application of η and after determining the value of $\eta(y))$ the operation ω is applied to its arguments. In this case, a function can be defined as having as domain the cartesian product of the ranges of x and y and, as codomain, the values assumed by $\omega(x, \eta(y)))$ in correspondence to the values of x and y:

$$(x, y) \mapsto \omega(x, \eta(y))).$$

The deep understanding of the relationship among arithmetic, logic, and computability is one of the most exciting successes of mathematical logic in the 20th century, explaining the pervasive role of symbolic sequences in computation processes [5, 6]. In a sense, molecular biology discovered the same kind of pervasiveness for biopolymers, which are representations of symbolic sequences at a molecular level.

The notion of expression can be mathematically generalized by the notion of a rooted tree, which expresses the abstract notion underlying a genealogical tree. A rooted tree is essentially obtained by starting from an element, called root, by applying to it a generation operation providing a set of nodes, that are the sons of the root, and then by iteratively applying to generated nodes the generation operation. When no generation operation is applied to a node, then it is a leaf. An element that is not a leaf is an internal node. A sequence of elements where any element (apart from the first) is the son of the previous one provides a branch of the tree, and a tree is infinite if it has an infinite branch.

Trees can be further generalized by graphs. A graph is given by a set A of elements, called nodes, and a set E of edges such that each edge is associated with a source node and a target node. A path in a graph is a sequence of nodes where for each node a of the sequence, apart from the last element of the sequence, exists another node b such that a and b are the source and the target of an edge of the graph. A path is infinite if no last element is present in the sequence, and it is a cycle if the last node of the path is equal to its first node. A graph is connected if, for any two nodes of the graph, there exists a path starting from one node and ending in the other one. A graph that is connected and without cycles is called a non-rooted tree.

Basic combinatorial schemata

Combinatorics deals with problems of counting finite structures of interest in many mathematical contexts [7]. These structures are usually definable in terms of sets, sequences, functions, and distributions over finite sets. Counting them is often not only mathematically interesting in itself, but very important from the point of view of a lot of applications. Combinatorics was the starting point of probability because some events correspond to the occurrence of structures of a given subclass within a wider class of possible structures. In many cases, a finite structure of interest is given by a **distribution** of objects into cells.

The number of functions from n objects to m cells is n^n (any object can be allocated in one of the cells), whereas the number of bijective functions from n objects to n, called also n-**permutations**, is obtained by allocating the first object in n ways, the second one in $n - 1$ ways (because cannot be allocated in the first cell), and so forth, with the last object allocated in only one possible way (in the last remained cell). Therefore:

$$|n \leftrightarrow n| = n!.$$

where $n!$ is the factorial of n, that is, the product $1 \times 2 \times \cdots \times (n-1) \times n$.

Reasoning as in the previous cases the number of injective functions $|n \hookrightarrow m|$ from n to m objects ($n \geq m$) is given by:

$$|n \hookrightarrow m| = n!_m$$

where $n!_m$ is the **falling factorial** of m decreasing factors starting from n:

$$n!_m = n(n-1)(n-2)\ldots(n-m+1) = \prod_{k=1}^{m}(n-k+1)$$

Counting the number of m-subsets of n elements can be obtained by counting all the injective functions $n \rightarrow_{\supset} m$ and then ignoring the distinguishability of cells, by dividing by $m!$, which corresponds to all the possible ways of ordering cells. In conclusion, the number $|m \subseteq n|$ of m-sets is given by:

$$|m \subseteq n| = n!_m/m!.$$

The m-sets over n objects are called **combinations** of m over n, and their numbers $n!_m/m!$ are called **binomial coefficients**, and denoted by:

$$\binom{n}{m} = n!_m/m!$$

also given by the following formula:

$$\binom{n}{m} = \frac{n!}{m!(n-m)!}.$$

The number of **partitions of** n **undistinguishable objects** in k distinct classes, possibly empty, can be obtained by considering boolean sequences of n zeros and $k - 1$ ones. A boolean sequence of $n + k - 1$ positions is completely determined when we chose the n positions where put zeros, therefore:

$$\binom{n+k-1}{n} = (n+k-1)_{!n}/n!$$

If we use the **rising factorial** notation $k!^n$ (the product of n increasing factors starting from k) it is easy to realize that:

$$\binom{n+k-1}{n} = k!^n/n!$$

If we choose k objects, over n distinct objects, with the possibility that an object is chosen many times, the choice can be represented by a boolean sequence of k zeros and $n-1$ ones, where zeros between two consecutive positions i and $i+1$ (also before the first one, or after the last one) express the multiplicity of the object of type i. A boolean sequence of $n+k-1$ positions is completely determined when we chose the k positions where put zeros, therefore the number of these choices is given by (the roles of objects and cells are exchanged, with respect to partitions):

$$\binom{n+k-1}{k} = n!^k/k!$$

The following proposition is an important characterization of binomial coefficients. Its proof follows a general schema, which is common to a great number of combinatorial formulas where the cases to count are partitioned into two distinct subsets.

Proposition 2.1 *The binomial coefficient can be computed by means of the Tartaglia-Pascal equation:*

$$\binom{n+1}{k+1} = \binom{n}{k} + \binom{n}{k+1}$$

Proof Let us consider an object, denoted by a_0, among the given $n+1$ objects. Then $k+1$-sets of $n+1$ elements can include a_0 or not. Therefore, we can count separately the number $C_{k+1,n}(a_0)$ of $k+1$-sets of $n+1$ elements including a_0 and the number $C_{k+1,n}(\neg a_0)$ of $k+1$-sets of $n+1$ elements not including a_0. The number $C_{k+1,n}(a_0)$ is given by $\binom{n}{k}$, because being a_0 included, any $k+1$-set corresponds to the choice of a k-set over n elements. The number $C_{k+1,n}(\neg a_0)$ is equal to $\binom{n}{k+1}$, because being a_0 not included, it can be removed from the set of $n+1$ objects, which becomes an n set, giving that $C_{k+1,n}(\neg a_0) = \binom{n}{k+1}$. In conclusion, the number of $\binom{n+1}{k+1}$ is the sum of $\binom{n}{k}$ and $\binom{n}{k+1}$.

2.2 Strings and Rewriting

Strings are the mathematical concept corresponding to the notion of words as linear arrangements of symbols. Dictionaries are finite sets of strings. These two notions are very important in the analysis of strings, and present a very rich catalogue of concepts and relations, in a great variety of situations.

Let us recall basic concepts and notation on strings (for classical textbooks, see for example [8]). Strings are finite sequences over an alphabet.

Often strings are written as words, that is, without parentheses and commas or spaces between symbols: $a_1a_2 \dots a_n$. When single elements are not relevant, generic strings are often denoted by Greek letters (possibly with subscripts). In particular, λ denotes the empty string. The length of a string α is denoted by $|\alpha|$ ($|\lambda| = 0$), whereas $\alpha[i]$ denotes the symbol occurring at position i of α, and $\alpha[i, j]$ denotes the substring of α starting at position i and ending at position j (the symbols between these positions are in the order they have in α).

A string can be also seen as a distribution when any symbol of the alphabet is associated with all the positions where it occurs in the string. Given an ordering $a < b < \cdots < z$ over the symbols of a string, the ordered normalization of a string α is the string

$$a^{|\alpha|_a} b^{|\alpha|_b} \dots z^{|\alpha|_z}$$

where $|\alpha|_x$ is the number of times symbol x occurs in string α and exponents denote the concatenation of copies of the symbol (a number of copies equal to the value of the exponent).

The set of all substrings of a given string α is defined by:

$$\text{sub}(\alpha) = \{\alpha[i, j] | 1 \leq i \leq j \leq |\alpha|\}.$$

and prefixes and suffixes are given by:

$$\text{pref}(\alpha) = \{\alpha[1, j] | 1 \leq j \leq |\alpha|\}$$

$$\text{suff}(\alpha) = \{\alpha[i, |\alpha|] | 1 \leq i \leq |\alpha|\}.$$

The most important operation over strings are:

1. **concatenation** of α, β, usually denoted by the juxtaposition $\alpha\beta$ The **overlap concatenation** of two strings $\alpha\gamma, \gamma\beta$, is $\alpha\gamma\beta$ where γ is the maximum substring that is the suffix of the first string and the prefix of the second one;
2. **length** usually denoted by the absolute value notation $|\alpha|$;
3. **prefix** $\alpha[1, j]$, with $1 \leq j \leq |\alpha|$;
4. **suffix** $\alpha[j, |\alpha|]$, with $1 \leq j \leq |\alpha|$;
5. **substitution** $subst_f$ with respect to a function f from the alphabet of α to another (possibly the same) alphabet, where $subst_f(\alpha)[j] = f(\alpha[j])$ for $0 < j \leq |\alpha|$;
6. **reverse** $rev(\alpha)$, such that $rev(\alpha)[j] = \alpha[n - j + 1]$;
7. **empty string** λ, such that, for every string α: $\lambda\alpha = \alpha\lambda = \alpha$. This string is very useful in expressing properties of strings and can be seen as a constant or nullary operation.

We remark the difference between **substring** and **subsequence**. Both notions select symbols occurring in a sequence in the order they are, but a subsequence does not assume that the chosen symbols are contiguous, while a substring of α is always of type $\alpha[i, j]$ and selects all the symbols occurring in α between the two specified positions (included). The number of all possible substrings of a string of length n has a quadratic order, whereas the number of all possible subsequences is 2^n.

A class of strings over an alphabet A is called a (formal) **language** (over the alphabet). Many operations are defined over languages: the set-theoretic operations (union, intersection, difference, cartesian product) and more specific operations as concatenation $L_1 L_2$ (of two languages), iteration L^n ($n \in \mathbb{N}$), and Kleene star L^*:

$$L_1 L_2 = \{\alpha\beta \mid \alpha \in L_1, \beta \in L_2\}$$
$$L^0 = \{\lambda\}$$
$$L^{n+1} = LL^n$$
$$L^* = \bigcup_{i \in \mathbb{N}} L^i.$$

Algorithms generating languages are called **grammars**. Automata are algorithms recognizing languages or computing functions over strings. Grammars are intrinsically non-deterministic while recognizing automata can be deterministic or non-deterministic. In the second case at each step many different consecutive steps are possible, and one of them is non-deterministically chosen. A string is recognized by such automata if for one of the possible computations the string is recognized.

We recall that a **Turing Machine** M [9] can be completely described by means of rewriting rules that specify as a tape configuration can be transformed into another tape configuration. A Turing configuration is a string of type:

$$\alpha q x \beta$$

where α is the string on the left of current position of M tape, β is the string on the right of this position, q is the current state, and x is the symbol of the tape cell that M is reading. In this way, an instruction of M such as $qxyq'R$ (in the state q and reading x rewrite x as y and pass to state q' moving on the cell to the right of the current cell) can be expressed by the following rewriting rule on configurations:

$$\alpha q x \beta \rightarrow \alpha y q' \beta$$

whereas an instruction of M such as $qxyq'L$ (in the state q and reading x rewrite x as y and pass to state q' moving on the cell to the left of the current cell) can be expressed by the following two rewriting rules on configurations:

$$\alpha q x \beta \rightarrow \alpha q'_* y \beta$$

$$\alpha x q'_* \beta \rightarrow \alpha q' x \beta.$$

The representation above of a Turing machine implies a very important property of the string rewriting mechanism: any process of string rewriting can be expressed by means of rules involving at most two consecutive symbols. In fact, the rules of the representation above involve the symbol of the state and only one symbol to the right or to the left of it. Moreover, according to the Church-Turing thesis, any computation can be realized by a suitable Turing machine, therefore any computation can be described by 2-symbol rewriting rules. This is a well-known result in formal language theory (Kuroda normal form representation).

Universal Turing machines U exist that verify the following universality equation:

$$U(< M, \alpha >) = M(\alpha)$$

where $M(\alpha)$ denotes the computation of Turing machine M with the input string α. The angular parentheses $< >$ denote an encoding of the pair given by machine M (its instructions) and input string α. Equality means that the left computation halts if and only if the right computation halts, and moreover, their final configurations are essentially equal (being in a 1-to-1 correspondence). In 1956 [10] Shannon found a universal Turing machine [9, 11] working with only two symbols, say 1 and B (the symbol B for blank).

An important class of grammars are the **Chomsky Grammars** essentially given by a set of rewriting rules $\alpha \rightarrow \beta$, where $\alpha \in A^* - T^*$, $\beta \in A^*$, $S \in A - T$, $T \subset A$ (T is called the terminal alphabet), and $\Rightarrow_*$ is the rewriting relation between a string and any other string obtained by a chain of substring replacements where a left side of a rule of a grammar G is replaced by its corresponding right side. The language determined by G is given by:

$$L(G) = \{\gamma \in T^* | S \Rightarrow_* \gamma\}.$$

A Chomsky grammar is of type 0 if no restriction is given to its rules. Chomsky grammars of type 0 generate all the recursively enumerable (or semi-decidable) languages, which belong to the class RE. For any language $L \in RE$ an algorithm exists that provides a positive answer to the question $\alpha \in L$ if and only if $\alpha \in L$. The class REC of recursive (or decidable) languages is a proper subclass of RE such that, for any $L \in REC$ an algorithm exists that provides a positive answer to the question $\alpha \in L$ if $\alpha \in L$ and a negative answer if $\alpha \notin L$ (proper inclusion $REC \subset RE$ was one of the most important achievements of the epochal paper of Alan Turing, in 1936 [9]).

Recursive enumerable languages coincide with the domains or codomains of Turing's computable functions.

2.3 Variables and Distributions

Given a variable X, we denote by $\widehat{X}$ its set of variability or its range. It is important to remark that the set $\widehat{X}$ of elements where X takes values can be completely unknown, or not relevant.

When a probability is associated to a set of values assumed by variable X, then X is a random variable, which identifies a probability distribution, where the whole unitary probability is distributed among the subsets of $\widehat{X}$.

The notion of distribution is very general and is a key concept in mathematical analysis, measure theory, probability, statistics, and information theory. Intuitively, a distribution specifies as a quantity is distributed in parts among the points or the regions of a space. A very simple case of a **finite distribution** is a multiset of objects where a number m of occurrences is distributed among k objects where n_1 are the occurrences of the first object, n_2 those of the second one, and so on, up to n_k, the occurrences of the last object, and:

$$\sum_{i=1...k} n_i = m.$$

A discrete probability distribution is a family of real numbers p_i, indexed in a set I such that:

$$\sum_{i \in I} p_i = 1.$$

2.4 Events and Probability

The theory of probability emerged in the 17th century, with some anticipations by the Italian mathematician Girolamo Cardano (1501–1576) (*Liber de ludo aleae*), Galileo (1564–1642) (*Sopra le scoperte dei dadi*, that is, About discoveries on dice), and Christian Huygens (1629–1695) (*De ratiociniis in ludo aleae*). The basic rules for computing probabilities were developed by Pascal (1623–1662) and Fermat (1601–1665). The first treatise on this subject was the book *Ars Conjectandi* by Jacob Bernoulli (1654–1705), where binomial coefficients appear in the probability of urn extractions, and the first law of large numbers is enunciated (*Theorema Aureus*: frequencies approximate to probabilities in a large space of events). The idea of events to which degrees of possibility are assigned is a change of perspective where facts, as they are (*modus essendi*), are replaced by facts as they are judged (*modus conjectandi*).

After Bernoulli's work, De Moivre (1667–1754) and Laplace (1749–1827), found the normal distribution as a curve for approximating the binomial distribution.

Bayes (1702–1752) discovered the theorem named by his name (independently discovered also by Laplace) ruling the *inversion of conditional probabilities*. Gauss

(1777–1855) recognized the normal distribution as the law of error distribution, and Poisson (1781–1840) introduced his distribution ruling the law of rare events.

An **event** can be expressed by means of a proposition asserting that a variable X assumes a value belonging to a given subset. This means that the usual propositional operations $\neg$, $\wedge$, $\vee$ (*not, and, or*) are defined on events. A space of events is a special boolean algebra (with 0, 1, sum, product, and negation)

A probability measure can be easily assigned to an event, as a value in the real interval [0, 1], which can be seen as a sort of evaluation of the possibility that the event has of happening.

The theory of probability concerns two different aspects:

(i) the study of probabilistic laws held in spaces of events,
(ii) the determination of the space of events which is more appropriate in a given context.

The first aspect constitutes a conceptual framework which is often independent of the specific spaces of events. A comparison may help to distinguish the two aspects. Calculus and the theory of differential equations provide rules and methods for solving and analyzing differential equations, but the choice of the right equation which is the best description of a physical phenomenon is a different thing, which pertains to the ability to correctly model phenomena of a certain type.

The axiomatic approach in probability theory was initiated by Andrej Nikolaevič Kolmogorov (1903–1987) and was developed by the Russian mathematical school (Andrey Markov, Pafnuty Chebyshev, Aleksandr Yakovlevich Khinchin). It is comparable to the axiomatic approach in geometry, and it is important for understanding the logical basis of probability. From the mathematical point of view, probability theory is part of a general field of mathematics, referred to as *Measure theory*, initiated by French mathematicians of the last century (Émile Borel, Henri Lebesgue, Nikolai Luzin, Johann Radon, Constantin Carathéodory, and Maurice Fréchet).

Basic rules of probability

The **conditional probability** of an event A, given an event B, is denoted by $P(A|B)$. It expresses the probability of A under the assumption that event B has occurred. Formally (propositional connectives or set-theoretic operations are used on events):

$$P(A|B) = \frac{P(A \wedge B)}{P(B)}.$$

Events A and B are said to be **independent**, and we write $A||B$, if $P(A|B) = P(A)$. Events A and B are **disjoint** if $P(A \wedge B) = 0$. The following rules connect propositional operations to probabilities. Proposition $\neg A$ has to be considered in terms of complementary set, that is if $A = (X \in S)$, then $\neg A = (X \in \widehat{X} - S)$.

1. $P(A) \geq 0$
2. $P(A \vee \neg A) = 1$
3. $P(A \wedge \neg A) = 0$
4. $P(\neg A) = 1 - P(A)$
5. $P(A \vee B) = P(A) + P(B) - P(A \wedge B)$
6. $P(A|B) = P(A \wedge B)/P(B)$
7. $A||B \Leftrightarrow P(A \wedge B) = P(A)P(B)$.

The theory of probability is the field where even professional mathematicians can be easily wrong, and very often reasoning under probabilistic hypotheses is very slippery. Let us consider the following examples from [12]. A military pilot has a 2% chance of being shot down in each military flight. What is the probability to die in fifty flights? We might guess that he is almost sure to die. But this is completely wrong. The probability of dying is the sum of the probabilities of dying at the first, at the second, and so on, up to the probability of dying at the last mission. Let us denote by p the probability of dying, then the sum of probabilities in all the flights is (the flight i is possible only if the pilot survives in the preceding flights):

$$p + (1 - p)p + (1 - p)^2 p + \ldots + (1 - p)^{49} p$$

A shorter evaluation of this probability (completely equivalent to the equation above) is the probability of surviving 0.98^{50} subtracted to 1. Therefore, the probability of dying is $1 - 0.98^{50} = 0.64$, which is very different from 1.

A similar error was the origin of the problem posed by the *Chevalier de Méré* to Pascal, about a popular dice game: *Why the probability of one ace, rolling one die 4 times, is greater than that of both aces, rolling two dice 24 times?* In fact, the probability of one ace is $1/6$, and $4/6 = 2/3$, analogously the probability of 2 aces is $1/36$, and $24/36 = 2/3$, so we may suppose that the two events: "1 ace in 4 rolls", and "2 aces in 24 double rolls" are equiprobable. But the empirical evidence reported by Chevalier de Méré was against this conclusion. Namely, Pascal (discussing with Fermat) solved the apparent paradox. In the first game, P(no-ace-in-4-tolls) = $(5/6)^4$, therefore P(ace-in-4-rolls)= $1 - (5/6)^4 = 0.5177$. In the second game, P(no-2-aces-in-24-double-rolls)= $(35/36)^{24}$, therefore P(2-aces-in-24-double-rolls) = $1 - (35/36)^{24} = 0.4914$. The simple mistake, as in the case of the military pilot, is due to the sum of probabilities that are not disjoint (by ignoring rule 5 given above).

Bayes' theorem

Cases of wrong probability evaluations are very frequent in contexts where conditional probabilities are present. Bayes' theorem, explains the nature of this difficulty because it establishes the rule for inverting conditional probabilities.

In a simplified form, Bayes' theorem asserts the following equation:

Proposition 2.2

$$P(A|B) = P(A)P(B|A)/P(B).$$

Proof By the definition of conditional probability, we have:

$$P(A \wedge B) = P(A|B)P(B)$$

analogously:

$$P(A \wedge B) = P(B \wedge A) = P(B|A)P(A)$$

therefore, by equating the right members of the two equations above we obtain:

$$P(A|B)P(B) = P(B|A)P(A)$$

from which the statement claimed by the theorem easily follows. □

Despite the simplicity of this proof, what the theorem asserts is an inversion of conditions. In fact, $P(A|B)$ can be computed by means of $P(B|A)$. Assume that a test T for a disease D is wrong only in one out of 1000 cases. When a person is positive on this test, what is the probability of having D? Is it $1 - 0.001 = 0.999$? No. In fact, this value confuses $P(D|T)$ with $P(T|D)$. The right way to reason is the application of Bayes' theorem; let us assume to know that D affects one out of 10000 persons, so $P(D) = 1/10000$, and that T is very reliable, with $P(T|D) = 1$. The probability $P(T)$ is $11/10000$ because T is wrong from 1 to 1000, and in only one out of 10000 persons with the disease D, test T is positive. In conclusion, the probability of having D is less than 10%. In fact:

$$P(D|T) = P(D)P(T|D)/P(T) = 1/10000 \times 1/(11/10000) = 1/11.$$

Statistical indexes and Chebicev's inequality

Let μ the **mean** of a random variable X assuming real values $x_1, x_2, \ldots x_n$ with probabilities $p_1, p_2, \ldots p_n$ respectively, then μ, denoted also by $E(X)$, is defined by the following equation:

$$E(X) = \sum_i p_i x_i$$

the second order moment of X is given by:

$$E(X^2) = \sum_i p_i x_i^2$$

and in an analogous way, moments of higher orders can be defined. the **variance** of X is given by the second-order momentum of the deviation from the mean, that is:

$$Var(X) = E[(X - \mu)^2)] = \sum_i p_i(x_i - \mu)^2.$$

The following equation holds:

$$var(X) = E(X^2) - [E(X)]^2.$$

In fact, from definitions, we obtain that:

$$\sum_i p_i(x_i - \mu)^2 = \sum_i p_i x_i^2 + \sum_i p_i \mu^2 - 2\mu \sum_i p_i x_i$$
$$= E(X^2) + \mu^2 - 2\mu^2 = E(X^2) - [E(X)]^2.$$

The standard deviation of X, $sd(X)$ is defined by:

$$sd(X) = \sqrt{Var(X)}.$$

The following theorem states a basic constraint of any statistical distribution: *given a random variable X, the probability that X assume a value differing, in absolute value, more than t from the mean μ is less than t^{-2} times the variance of X.*

Proposition 2.3 (Chebichev Theorem)

$$P(|X - \mu| > t) \leq t^{-2} E[(X - \mu)^2]$$

Proof

$$P(|X| > t) \leq t^{-2} E(X^2)$$

In fact:

$$\sum_{|x| \geq t} p(x) \leq \sum_{|x| \geq t} p(x) \frac{x^2}{t^2} \leq t^{-2} E(X^2)$$

whence, replacing X by $(X - \mu)$ the thesis holds. □

A direct consequence of the theorem above is the so-called **Law of large numbers** in its weak form, originally proved by Bernoulli, stating that in a boolean sequence of successes and failures, the frequency of successes rends to the success probability according to which the sequence is generated. In formal terms. let $S(\alpha_n)$ be a boolean sequence generated by a n Bernoulli boolean variables $X_1, X_2, \ldots, X_n$ all with a probability of success (that is of having value 1) equal to p. Let $S(\alpha_n) = S_n$ the number of successes in α_n, $S_n = X_1 + X_2, \ldots, +X_n$ and Ω_n the set of all sequences of length n. The following limit holds.

Proposition 2.4

$$\forall \varepsilon \in \mathbb{R}^+ : \lim_{n\to\infty} P\{\alpha_n \in \Omega_n : |S_n/n - p| > \varepsilon\} = 0$$

Proof The mean of a Bernoulli boolean variable X is $E(X) = 1 \times p + 0 \times (1 - p) = p$. Therefore being S_n the sum of n independent variables $E(S_n) = np$. Analogously, the square deviation of any boolean random variable X is ($q = 1 - p$):

$$(1-p)^2 p + (0-p)^2 q = q^2 p + p^2 q = qp(q+p) = qp$$

therefore the sum of n independent boolean variables with success probability p has square deviation $npq = np(1 - p)$. Fix ε, then by Chebichev's inequality:

$$P\{\alpha_n \in \Omega_n : |\frac{S_n}{n} - E(S_n)| > \varepsilon\} \leq \frac{Var(S_n)}{\varepsilon^2}.$$

The mean value and the variance of $\frac{S_n}{n}$ are:

$$E\left(\frac{S_n}{n}\right) = \frac{E(S_n)}{n} = p$$

$$Var\left(\frac{S_n}{n}\right) = \frac{Var(S_n)}{n^2} = \frac{np(1-p)}{n^2} = \frac{p(1-p)}{n}$$

thus, we obtain:

$$P\{\alpha_n \in \Omega_n : |\frac{S_n}{n} - p| > \varepsilon\} \leq \frac{p(1-p)}{n\varepsilon^2}$$

so, passing to the limit:

$$\lim_{n\to\infty} P\{\alpha_n \in \Omega_n : |\frac{S_n}{n} - p| > \varepsilon\} \leq 0$$

but the probability cannot be negative, therefore:

$$\lim_{n\to\infty} P\{\alpha_n \in \Omega_n : |\frac{S_n}{n} - p| > \varepsilon\} = 0.$$ □

Besides, mean, momenta, variance, and standard deviation, the **correlation coefficient** is an important index for expressing the degree of reciprocal influence between two variables, on the basis of the distributions of their values. Correlation is based on the notion of **covariance** that is expressed by the mean of the product of deviations (μ and ν are the means of X and Y, respectively):

$$cov(X, Y) = E[(X - \mu)[Y - \nu)].$$

Then, the correlation coefficient $\rho(X, Y)$ is given by (assuming a finite variance for X and Y):

$$\rho(X, Y) = \frac{cov(X, Y)}{Var(X)Var(Y)}.$$

The intuition concerning covariance is very simple. If two variables deviate from their means in a coherent way they are related, and the product of the deviations is positive when their deviations agree and is negative otherwise. Moreover, the agreements and disagreements are reasonably weighted by the probabilities with which they occur. The denominator in the correlation ratio is introduced to normalize the covariance with respect to the variances of the two variables.

References

1. Manca, V.: Il Paradiso di Cantor. La costruzione del linguaggio matematico, Edizioni Nuova Cultura (2022)
2. Manca, V., Arithmoi. Racconto di numeri e concetti matematici fondamentali (to appear)
3. Fränkel, A.A.: Set Theory and Logic. Addison-Wesley Publishing Company (1966)
4. Halmos, P.: Axiomatic Set Theory. North-Holland (1964)
5. Manca, V.: Formal Logic. In: Webster, J.G. (ed.) Encyclopedia of Electrical and Electronics Engineering, vol. 7, pp. 675–687. Wiley, New York (1999)
6. Manca, V.: Logical string rewriting. Theor. Comput. Sci. **264**, 25–51 (2001)
7. Aigner, M.: Discrete Mathematics. American Mathematical Society, Providence, Rhode Island (2007)
8. Rozenberg, G., Salomaa, A.: Handbook of Formal Languages: Beyonds words, vol. 3. Springer, Berlin (1997)
9. Turing, A.M.: On computable numbers, with an application to the Entscheidungsproblem. Proc. Lond. Math. Soc. **42**(1), 230–265 (1936)
10. Shannon, C.E.: A universal Turing machine with two internal states. In: Automata Studies, Annals of Mathematics Studies, vol. 34, pp. 157–165. Princeton University Press (1956)
11. Minsky, M.L.: Computation: Finite and Infinite Machines. Prentice-Hall Inc. Englewood Cliffs, N. J. (1967)
12. Aczel, A.D.: Chance. Thunder's Mouth Press, New York (2004)

Chapter 3
Information Theory

Introduction

Information theory "officially" begins with Shannon's booklet [1] published in 1948. The main idea of Shannon is linking information with probability. In fact, the starting definition of this seminal work is that of **information source** as a pair (X, p), where X is a finite set of objects (data, signals, words) and p is a probability function assigning to every $x \in X$ the probability $p(x)$ of occurrence (emission, reception, production). The perspective of this approach is the mathematical analysis of communication processes, but its impact is completely general and expresses the probabilistic nature of the information. Information is an inverse function of probability. because it is a sort of a posteriori counterpart of the a priori uncertainty represented by probability, measuring the gain of knowledge when an event occurs. For this reason, the more an event is rare, the more it is informative. However, if event E has probability p_E, for several technical reason it is better to define $inf(E)$ as $\lg(1/p_E) = -\lg(p_E)$ rather than $1/p_E$. In fact, the logarithm guarantees the information additivity for a joint event (E, E') where components are independent, giving $inf(E, E') = inf(E) + inf(E')$. However, it is important to remark that the relationship between information and probability is in both verses because as Bayesian approaches make evident, information can change the probability evaluation (conditional probability, on which Bayes theorem is based, defines how probability changes when we know that an event occurred). A famous example of this phenomenon is the three-doors (or Monty Hall) dilemma [2], which can be fully explained by using the Bayes theorem.

In this chapter, we give a quick overview of the basic concepts in Information Theory. The reader is advised to refer to [1, 3, 4] for more details on Information and Probability theories. Some specific concepts and facts about probability will be more exhaustively considered in the next chapter.

V. Manca and V. Bonnici, *Infogenomics*, Emergence, Complexity and Computation 48,
https://doi.org/10.1007/978-3-031-44501-9_3

3.1 From Physical to Informational Entropy

Shannon founded Information Theory [1, 3] on the notion of **Information Source**, a pair (A, p), given by a set of data A and a probability distribution p defined on A, or even, a pair (X, p_X) of a variable X assuming values with an associated probability distribution p_X (A is the set on which X assumes values). In this case, only the variable X can be indicated if p_X is implicitly understood. In this framework, Shannon defined a measure of **information quantity** of the event $X = a$ (X assumes the value a):

$$Inf(X = a) = -\lg_2(p_X(a))$$

where $p_X(a)$ is the probability of the event $(X = a)$. The intuitive motivation for the equation above is that the information quantity associated with an event is inversely proportional to the probability of the event, and moreover, the information quantity has to be additive for pairs of independent events $(p(a_1, a_2) = p(a_1)p(a_2))$:

$$Inf(E_1, E_2) = Inf(E_1) + Inf(E_2).$$

On the basis of this definition, if $\widehat{X}$ denotes the range of the variable X, the **entropy** of the information source (X, p_X) is defined by:

$$H(X, p_X) = \sum_{a \in \widehat{X}} p_X(a) Inf(X = a) = -\sum_{a \in \widehat{X}} p_X(a) \lg_2 p_X(a).$$

Therefore, the entropy of an information source is the mean (in a probabilistic sense) of the information quantity of the events generated by the information source (X, p_X).

One crucial result about entropy is the **Equipartition Property** (proved by Shannon in an appendix to his booklet): in the class of variables assuming the same value of X the value of the entropy reaches its maximum with a source (Y, q_Y) where q_Y is the uniform probability distribution (all the values of variable Y are assumed with the same values).

Let us recall that Shannon's idea has very ancient historical roots (probably unknown to Shannon). The encoding method "Caesar" (used by Julius Caesar) is a very simple way of encoding messages for hiding their contents so that only whoever knows a deciphering key can access to them. This method is based on a one-to-one function assigning a letter $f(X)$ to a letter X. Given a text, if you replace each letter by using this function, then the message becomes unreadable unless you use the inverse of f for recovering the original message. In the eightieth century, the Arabic mathematician Al-Kindi discovered a method for cracking "Caesar" encoding, entirely based on frequencies. Let us assume to know the language of the original messages, and to collect a great number of encrypted messages. In any language, the frequency of each letter is almost constant (especially if it is evaluated for long texts). Then, if we compute the frequencies of letters in the encrypted messages, it results that, with a very good approximation, we can guess, for every letter of the alphabet which is the letter in the encrypted texts having a similar frequency. This letter is, with a good probability the letter corresponding to the letter of plain texts, in the encrypted messages. In this way, the deciphering key f can be discovered and messages can be disclosed.

The consequence of this deciphering method is that the essence of any symbol in a symbolic system is intrinsically related to its frequency, which is a special case of probability (an empirical probability). For this reason, information is directly related to probability.

The notion of informational entropy is strongly related to thermodynamic entropy S, which emerged in physics since Carnot's analysis of heat-work conversion [5] and was named by Clausius "entropy" (from Greek en-tropos meaning "internal verse"). Physical entropy is subjected to a famous inequality stating the **Second Principle of Thermodynamics** for isolated systems (systems that do not exchange energy with their environment), entropy cannot decrease in time:

$$\Delta S \geq 0.$$

In years 1870s, Ludwig Boltzmann started a rigorous mathematical analysis of a thermodynamic system consisting of ideal gas within a volume, aimed at explaining the apparent contradiction of the inequality above with respect to the equations of Newtonian mechanics, underlying the microscopic reality on which heat phenomena are based on. More precisely, Boltzmann's question was: "From where the inequality comes from, if molecules colliding in a gas follow mechanics equational laws with no intrinsic time arrow?"

For answering the above question, Boltzmann introduced, in a systematic way, a probabilistic perspective in the microscopic analysis of physical phenomena. For its importance, his approach transcends the particular field of his investigation. In fact, after Boltzmann, probability and statistics [4, 6, 7] became crucial aspects of any modern physical theory. Boltzmann defined a function H, which in discrete terms can be expressed by:

$$H = \sum_{i=1,n} n_i \lg_2 n_i$$

where n_i are the number of gas molecules having velocities in the i-class of velocities (velocities are partitioned in n disjoint intervals).

By simple algebraic passages, it turns out (details are given in Section 5.4) that the H function coincides, apart from additive and multiplicative constants, with Shannon's entropy (Shannon quotes Boltzmann's work). On the basis of H function Boltzmann proved the microscopic representation of Clausius entropy S [8].

$$S = k \log_e W \tag{3.1}$$

where W is the number of distinguishable micro-states associated with the thermodynamic macro-state of a given system (two micro-states are indistinguishable if the same velocity distribution is realized apart from the specific identities of single molecules) and k is the so-called *Boltzmann constant*.

However, despite this remarkable result, Boltzmann was not able to deduce that $\Delta S \geq 0$, by using H function. The so-called Theorem H stating $H_{t+1} \leq H_t$, or $\Delta H \leq 0$, was never proved by him in a satisfactory way (from $\Delta H \leq 0$ inequality $\Delta S \geq 0$ follows as an easy consequence of the above equation $S = k \log_e W$, where the opposite verses of inequalities are due to the different signs of Boltzmann's H and Shannon's H) [5, 9–12].

The microscopic interpretation of thermodynamic entropy given by Boltzmann has a very general nature expressed by the **Entropy Circular Principle** ECP, corresponding to Boltzmann's "Wahrscheinlichkeit" Principle. In terms of Shannon information sources, ECP can be formulated by the following statement.

The entropy of a source (X, p) is proportional to the logarithm of the number W of different sources generating the same probability distribution p, that is, the entropy of a given source is proportional to the number of sources having that entropy.

Entropy Circular Principle (ECP)

$$H(X, p_X) = c \lg_2 |\mathbb{X}|$$

for some constant c, where:

$$\mathbb{X} = \{(Y, p_Y) \mid \widehat{X} = \widehat{Y}\ ,\ p_X = p_Y\}$$

From this principle will emerge an interesting link between probabilistic and digital aspects of information. In fact, it can be proved that $\lg_2 n$ corresponds essentially to the minimum average number of binary digits necessary for encoding n different values (the exact value $digit_2(n)$ is bounded by: $(\lfloor lg_2(n) \rfloor - 2) < digit_2(n) < n(\lfloor lg_2(n) \rfloor)$). Therefore, entropy corresponds (apart from multiplicative constants) to the average length of codewords encoding all the sources with the same entropy. This digital reading of the entropy is also related to the First Shannon Theorem, stating that $H(X, p_X)$ is a lower bound for the (probabilistic) mean of yes-no questions necessary to guess a value $a \in \widehat{X}$ by asking $a \leq b$?, for suitable values b in $\widehat{X}$. This explains in a very intuitive manner the meaning of $H(X, p)$, as the average uncertainty of the source (X, p).

In this sense, entropy is, at the same time, the average uncertainty of the events generated by the source, and at the sane time, the average uncertainty in identifying the source, in the class of those with the same probability distribution. The uncertainty *internal* to the space of events coincides with the *external* uncertainty of the event space, among all possible event spaces.

In this perspective, the ECP principle unveils the intrinsic *circularity* of entropy, because $H(X, p_X)$ is completely determined by p_X but, at the same time, corresponds to the (logarithm of the number of) ways of realizing distribution p_X by random variables assuming the same values of X.

The reason for Boltzmann's failure in correctly proving H theorem is due to the fact that this theorem is, in its essence, an information Theory theorem (see [8,

12] for proofs of H theorem). In fact, in simple words, the irreversible increase of entropy in time, results from a collective effect of the large number of reversible elementary events, according to the casual nature of molecular collisions. Molecules that collide, exchanging information, produce, in time, an increasing uniformity of molecular velocities. The global energy does not change (collisions are elastic), but differences between velocities decrease and this provides a greater value of the entropy, according to the entropy equipartition property.

3.2 Entropy and Computation

We showed how information entropy is related to physics and how time in physics is related to entropy. In computations the informational and physical components are intertwined, because any computational device, even when abstractly considered, is based on a physical process, where states, symbols, or configurations are transformed along a dynamics starting with some initial configurations, and eventually ending, if computation reaches a result, with a final configuration (the initial configuration encodes input data, while the final configuration encodes the corresponding output data).

In this perspective, computation is a trajectory in space of events, where events can be conceived as suitable sets of states (according to micro-state/macro-state Boltzmann's distinction, any macro-state consists of a collection of micro-states).

On the other side, if computation is a process intended to acquire information, this representation suggests that the configurations along a computation have to reduce the number of internal micro-states. In fact, the uncertainty of a configuration corresponds with the number of states it contains, and a computation tends to reduce the initial uncertainty by reaching more certain configurations, which provide a solution, by satisfying the constraints of the given problem.

In terms of entropy, if the number of possible states reduces along a computation, then the entropy $\lg W$ (W number of states) decreases along the computation. In conclusion, according to the outlined perspective, the computation results to be an anti-entropic process. For this reason, from a physical point of view, a computation is a dissipative process releasing energy (heat) in the environment. A computational device is comparable with a living organism: it increases the environment entropy for maintaining his low internal entropy [11], analogously, a computational device increases the environment entropy for proceeding in the computation and providing a final result.

An intuitive way for realizing that "information is physical" is a simple device called *Centrifugal Governor*, invented by James Watt in 1788 for controlling his steam engine. The principle on which it is based is that of a "conical pendulum". The axis of a rotating engine is supplied with an attached rigid arm terminating with a mass that can assume an angle with respect to the axis (two symmetrical arms in Watt's formulation, with a negligible mass at end of any arm). When the axis is rotating, in proportion to the rotation speed, the mass at end of the arm is subjected to a centrifugal

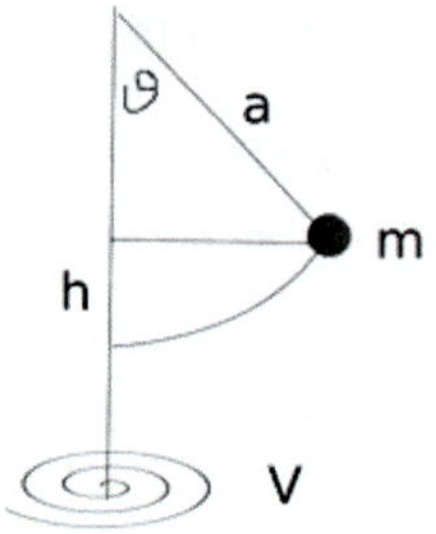

Fig. 3.1 The schema of *Centrifugal Governor*

force rising the arm (see Fig. 3.1). This rising opens a valve, in proportion to the vertical rising of the arm, which diminishes a physical parameter related to the speed (pressure in the case of a steam engine), and consequently, decreases the rotation speed. This phenomenon is a **negative feedback** that, according to the length of the arm, stabilizes the rotation speed to a fixed velocity (negative means that control acts in the opposite verse of the controlled action). It realizes a kind of **homeostasis** (keeping some variables within a fixed sub-range of variability), a property that is crucial for any complex system and thus typical of living organisms.

The term "cybernetics" introduced by Norbert Wiener [13] comes from a Greek root expressing the action of guiding or controlling, and is essentially based on the image of a Centrifugal Governor, where the arm length encodes a piece of information capable of controlling the rotation speed (Wiener's book title is: Cybernetics, or Control and Communication in the Animal and the Machine). Information directs processes, but it is realized by a physical process too. And in the centrifugal governor, mgh is the energetic cost of the control exerted by the arm on the rotation (see Fig. 3.1: m is the mass at the end of the arm, g the gravity acceleration, and h is the vertical rising of the arm). We remark that from equation $mv^2/2 = mgh$ (kinetic energy = potential energy) it follows that rising $h = v^2/2g$ does not depend on the value of the mass, but only on the speed (that has to be sufficient to provide a centrifugal force able to open the valve).

The investigation of the relationship between information and physics has a long history, going back to the famous *Maxwell demon*, introduced by the great physicist in addressing the problem of how an intelligent agent can interfere with physical principles. His demon was apparently violating the second principle of thermodynamics, for his ability to get information about the molecules' velocity (but also this acquisition of information requires an energetic cost). A very intensive and active line of research was developed in this regard (see [14] for a historical account of these researches) and continues to be developed, even in recent years, especially under the pressure of the new results in quantum information and in quantum computing, which show how much it is essential the informational perspective for a deep analysis of quantum phenomena [15].

It is becoming always more evident that any mathematical model of physical phenomena is based on data coming from information sources generated by measurement processes. This fact is not of secondary relevance, with respect to the construction

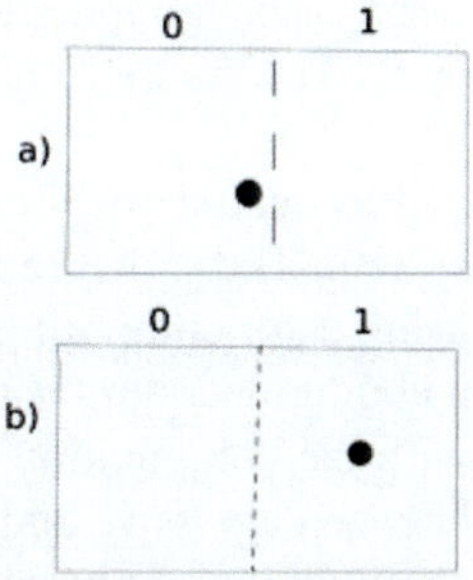

Fig. 3.2 Landauer's minimal erasure as a compression of the state space: in **a** the ball can be located in any of the two parts 0, 1 of the volume; in **b** it is confined in part 1, therefore in (**b**) the state space is reduced from two possibilities to only one of them

of the model, especially when the data acquisition processes are very sophisticated and cannot ingenuously be considered as mirrors of reality, rather, the only reality on which reconstructions can be based is just the information source resulting from the interactions between the observer and the observed phenomena.

In a celebrated paper [16] Rolf Landauer asserts the *Erasure Principle*, according to which the erasure of a bit during a computation has an energetic cost of $kT \ln 2$ (k Boltzmann constant, T absolute temperature) [17]. This means that, if the computation entropy diminishes (for the reduction of computation configuration states), then the environment entropy has to compensate this decrease by releasing a corresponding quantity of heat ST in the environment.

The erasure principle is nothing else than a direct derivation of Boltzmann Equation 3.1, when from two possible states one of them is chosen, by passing from $W = 2$ to $W = 1$. However, it is important to remark that "erasing" has to be considered in the wide sense of decreasing states of a computation configuration, as indicated in Fig. 3.2 (the term erasure could be misleading, because it means writing a special symbol on a Turing machine tape).

A continuation of Landauer's research has been developed by Bennet and others [14, 18–20] (in [14] a historical account of the theory of reversible computation is given). Computation steps can be reversible and irreversible. The first kind of step arise when information is not lost after the step and it is possible to go back by returning to the data previously available. The irreversible steps are those that do not satisfy the condition of reversibility because some data are lost after the step. In this framework, it is shown that any computation can be performed in a reversible way, by using suitable strategies, where all the steps of a computation ending in a given configuration are copied in a suitable zone of memory such that from the final state it is possible to go back in a reverse way. Of course, in this way computation is reversible, but we are perplexed about the effective meaning of this result. In fact, if the information is gained by reducing the set of states of the initial configuration, then in the case of a reversible computation no new information is obtained at end of the computation. This would require a different way of considering computations, where the evaluation of the amount of the gained information, at the end of computation, has to be defined in other terms. Otherwise, if a computation does not generate information, for what reason it is performed? What is the advantage of obtaining a result when computation halts? If a result is not new information, what gain is given by it? The theory of reversible computation, even if correct, is surely incomplete, as far as it does not answer these open questions [21].

Physical states representing data are a proper subset of all physical states of a system. Those that represent data are selected by the observer of the system who is using it for computing. But computation is not only a physical activity, it is mainly a coding-decoding activity according to a code chosen by another system coupled with the computation. This means that in any computation we need to distinguish between a passive and an active component. Forgetting this coupling and analyzing computations only on the side of the operations applied to the tape (of a Turing machine) can be useful, but does not tell the whole story of the computing process. This partiality is the source of some conclusions drawn in the theory of reversible computation. If even all the logical gates are reversible, the internal states of the "interpreter" processor could forget an essential aspect of the computation. Therefore, is not so obvious that we can always get rid of erasing operations, when also the program determination is taken into account, rather than only data transformations.

The hidden trap, in analyzing a computation, is that of considering only the entropy of data, by forgetting that an agent provides the input data and a program, and he/she decodes the data obtained at end of the computation (if it halts). In other words, no computation can forget the agent performing it. Computation is not only in the computing device, but also in the interaction between an agent and a computation device. For this reason, also the entropic/energetic evaluation of the process that in the agent is coupled with the computing device has to be taken into account.

For example, in reversible computation theory, a comparison is often presented with the RNA Polymerase transcription from DNA to RNA, by arguing that this copy phenomenon is reversible. But in this comparison, no mention is done about the anti-entropic character of RNA Polymerase that, in order to be maintained "alive", requires some corresponding anti-entropic processes.

In conclusion, the theory of reversible computation is sure of great interest, by disclosing subtle relations among time, space, and energy [14], however, what it claims raises several controversial aspects that need further clarifications and experiments.

An anecdote reports that when Shannon asked John von Neumann to suggest him a name for the quantity S, then von Neumann promptly answered: "Entropy. This is just entropy", by adding that this name would have been successful because only a few men knew exactly what entropy was.

Entropy has a paradoxical aspect, due to its intrinsic probabilistic nature. The paradox is apparent at the beginning of Shannon's paper [1]. In fact, Shannon says that he is searching for a measure of information, or uncertainty. But how can we consider these notions as equivalent? It would be something like searching for a measure of knowledge or ignorance. Can we reasonably define the same measures for opposite concepts? The answer to these questions can be found in the intrinsic orientation of events in time. When an event e happens with a probability p, we can measure its information by a function of its a priori uncertainty p, but after it happens, we gain, a posteriori, the same amount of information associated to p, therefore a priory uncertainty is transformed into information. The same kind of opposite orientation is always present in all informational concepts and it is often a source of confusion when the perspective of considering them is not clearly defined [7].

As we will show, data can be always expressed by strings, that is, data can be linearized over a given alphabet of symbols. This assumption is a prerequisite of Shannon's analysis that can be summarized by his three fundamental theorems.

First Shannon theorem provides a lower bound to the mean length of strings representing the data emitted by an information source (mean calculated with respect to the probability distribution of the source). More precisely, the entropy of the source, which is independent of any method of data representation, turns out to provide this lower bound.

Second Shannon theorem is based on mutual information: *even if data of an information source are transmitted with some noise along a channel, it is possible to encode them in such a way that transmission could become error-free, because there exist transmission codes such that, the longer are the transmission encodings, the more error probability approaches to zero.*

In more precise terms, the theorem establishes quantitative notions giving the possibility of avoiding error transmission when the **transmission rate** is lower than the channel **capacity of transmission**, where these notions are formally defined in suitable terms by using mutual information (between the **transmitter** information source and the bf receiver information source).

Third Shannon theorem concerns with signals. To this end, the entropic notions are extended to the case of continuous information sources, and then, by using these continuous notions, quantitative evaluations are proven about safe communications by means of continuous signals.

3.3 Entropic Concepts

Entropy allows us to define other informational concepts further developing the probabilistic approach.

Let us remark that the notion of information source is completely equivalent to that of an aleatory or random variable, that is, a variable X to which is associated a probability p_X, where $p_X(x)$ is the probability that X assumes the value x. In other words, the random variable X identifies the information source (A, p_X), where A is the set of values assumed by the variable X. *Viceversa*, an information source (A, p) identifies a random variable X that assumes the values of A with the probability distribution p. For this reason, very often, entropy and other entropic concepts, are equivalently given for information sources, random variables, and/or probability distributions (discrete or continuous). In the following, according to the specific context, we adopt one of these alternatives.

Conditional entropy

Given two random variables X e Y of corresponding distributions p_X and p_Y, the joint variable (X, Y) is that assuming as values the pairs (x, y) of values assumed by X and Y respectively, where p_{XY} is the joint distribution $p_{XY}(x, y)$, shortly indicated by $p(x, y)$, giving the probability of the composite event $X = x$ and $Y = y$ (when arguable subscripts in p_X, p_Y, p_{XY} will be avoided).

The **joint entropy** $H(X, Y)$ of (X, Y) is defined by:

$$H(X, Y) = -\sum_{x \in A} \sum_{y \in B} p(x, y) \log p(x, y)$$

or:

$$H(X, Y) = -\sum_{x, y \in A \times B} p(x, y) \log p(x, y).$$

By definition of conditioned probability, we have that $p(y|x) = \frac{p(x,y)}{p(x)}$. Given two random variables X e Y, **conditional entropy** $H(Y|X)$ is defined by the following equation, where $E[\ldots]$ denotes the expected value with respect to the distribution probability of the random variable inside brackets:

$$H(Y|X) = -\sum_{x, y \in A \times B} p(x, y) \log p(y|x)$$

$$\begin{aligned} H(Y|X) &= -\sum_{x, y \in A \times B} [p(x)p(y|x)] \log p(y|x) \\ &= -\sum_{x \in A} p(x) \sum_{y \in B} p(y|x) \log p(y|x) \\ &= \sum_{x \in A} p(x) H(Y|X = x). \end{aligned}$$

In the above definition, the joint probability is multiplied by the logarithm of the conditional entropy (rather than, conditioned probability multiplied by its logarithm). This apparent asymmetry is motivated by the following equation, which follows from the definition above, where for any value x of X, H_x denotes the entropy of X conditioned by the event $X = x$:

$$H(Y|X) = \sum_{x \in A} p(x) H_x.$$

The following proposition corresponds to the well-known relation between joint and conditional probabilities: $p(x, y) = p(x)p(y|x)$.

Proposition 3.1

$$H(X, Y) = H(X) + H(Y|X).$$

Proof

$$H(X, Y) = -\sum_{x,y \in A \times B} p(x, y) \log p(x, y)$$

$$= -\sum_{x,y \in A \times B} p(x, y) \log(p(x)p(y|x))$$

$$= -\sum_{x,y \in A \times B} p(x, y) \log p(x) - \sum_{x,y \in A \times B} p(x, y) \log p(y|x)$$

$$= -\sum_{x,y \in A \times B} p(x, y) \log p(x) + H\,(Y|X)$$

$$= -\sum_{x \in A} \cdot \sum_{y \in B} (p\,(x, y) \log p(x)) + H\,(Y|X)$$

$$= -\sum_{x \in A} \left(\sum_{y \in B} p\,(x, y) \right) \log p(x) + H\,(Y|X)$$

$$= -\sum_{x \in A} p(x) \log p(x) + H\,(Y|X)$$

$$H(X, Y) = H(X) + H\,(Y|X)$$

Entropic divergence

Given two probability distributions p, q their **entropic divergence**, $D(p, q)$, also denoted by $KL(p, q)$ after Kullback and Leibler [3], is the probabilistic mean, with respect to distribution p, of the differences of information quantities associated to distribution p and q respectively:

$$D(p, q) = \sum p(x)[\lg p(x) - \lg q(x)] \tag{3.2}$$

or, indicating by E_p the mean with respect to the probability distribution p:

$$D(p, q) = E_p[\lg p(x) - \lg q(x)] \tag{3.3}$$

or, by using a property of logarithm:

$$D(p, q) = E_p\left[\log \frac{p(x)}{q(x)}\right]$$

When $q = p$, $D(p, p) = 0$. Moreover, equations $0 \log \frac{0}{q} = 0$ e $p \log \frac{p}{0} = \infty$ are usually assumed.

Mutual Information

Let p be a probability distribution defined over $X \times Y$, where p_X, p_Y are the probabilities of X, Y respectively, also called **marginal probabilities**. The **mutual information** $I(X, Y)$ is given by the entropic divergence between distributions $p_{(X,Y)}$ and $p_X \times p_Y$, where p_{XY} is the joint probability assigning to each pair (x, y) the probability $p(x, y)$ of the joint event $X = x, Y = y$, while $p_X \times p_Y$ is the product of the marginal probabilities, assigning to each pair (x, y) the product $p(x)p(y)$ of the two probabilities p_X, p_Y:

$$I(X, Y) = D(p_{XY}, p_X \times p_Y). \tag{3.4}$$

By the definition of $I(X, Y)$ we have:

$$I(X, Y) = \sum_{x \in X, y \in Y} p(x, y) \lg[p(x, y)/p(x)p(y)]. \tag{3.5}$$

therefore:

$$I(X, Y) = \sum_{x \in X, y \in Y} p(x, y)[\lg p(x, y)/p(x)] - \sum_{x \in X, y \in Y} p(x, y) \lg p(y)] \tag{3.6}$$

that is:

$$I(X, Y) = \left(\sum_{x \in X, y \in Y} p(x, y)[\lg p(y|x)]\right) - \sum_{y \in Y} p(y) \lg p(y)] \tag{3.7}$$

that can be written as:

$$I(X, Y) = H(Y) - H(Y|X). \tag{3.8}$$

If we consider the conditional entropy $H(Y|X)$ as the **mean conditional information** of Y given X, then Eq. (3.8) tells us that the mutual information between X and Y is the mean information of Y minus the mean conditional information of Y given X.

Proposition 3.2 $D(p, q) \geq 0$

Proof The following is a very simple inequality resulting from the graphic of logarithm function:

$$\ln x \leq x - 1$$

therefore:

$$\ln \frac{q_i}{p_i} \leq \frac{q_i}{p_i} - 1$$

multiplying by p_i and summing over i on both members:

$$\sum_i p_i \ln \frac{q_i}{p_i} \leq \sum_i p_i \left(\frac{q_i}{p_i} - 1\right)$$

$$\sum_i p_i \ln \frac{q_i}{p_i} \leq \sum_i q_i - \sum_i p_i$$

because $\sum_{i=1}^{m} p_i = \sum_{i=1}^{m} q_i = 1$ (p e q are probability distributions):

$$\sum_i p_i \ln \frac{q_i}{p_i} \leq 0$$

or, by passing to logarithms in base k:

$$\sum_i p_i \log_k \frac{q_i}{p_i} \leq 0$$

then, reversing the logarithm fraction with the consequent change of sign, the opposite inequality is obtained, that is:

$$\sum_i p_i \log_k \frac{p_i}{q_i} \geq 0$$

$$D(p, q) \geq 0$$

□

Proposition 3.3 • $I(X, Y) = I(X, Y)$ *(symmetry)*
• $I(X, Y) \geq 0$ *(non-negativity)*
• $I(Y, X) = H(X) + H(Y) - H(X, Y)$

Proof The first equation follows directly from the definition of mutual information, while the second one follows from non-negativity of Kullback-Leibler divergence. The third equation follows from equations $I(X, Y) = H(X) - H(X|Y)$ and

$I(Y, X) = H(Y) - H(Y|X)$, by replacing in them the conditional entropies as they result from $H(X|Y) = H(X) - H(X, Y)$ and $H(Y|X) = H(Y) - H(Y, X)$. □

Entropic divergence and mutual information provide two important properties of entropy.

Proposition 3.4 *The entropy of a random variable X reaches its maximum when X is uniformly distributed.*

Proof If n is the number of values assumed by X (discrete and finite case), then:

$$H(X) \leq \log n.$$

In fact, if X is uniformly distributed we have:

$$H(X) = -\sum_{1}^{n} \frac{1}{n} \log \frac{1}{n} = -\frac{n}{n} \log \frac{1}{n} = \log n$$

If u denotes the uniform probability distribution:

$$\begin{aligned} D(p, u) &= \sum_{x} p(x) \log \tfrac{p(x)}{u(x)} \\ &= \sum_{x} p(x) (\log p(x) - \log u(x)) \\ &= \sum_{x} p(x) \log p(x) - \sum_{x} p(x) \log u\,(x) \\ &= -H(X) + \log n \end{aligned}$$

therefore:

$$-\,H(X) - \log 1/n \geq 0$$

that is:

$$H(X) \leq \log n$$

□

Proposition 3.5

$$H(X|Y) \leq H(X).$$

Proof

$$I(X, Y) = H(X) - H(X|Y)$$

and:

$$D(p, q) \geq 0$$

therefore: $I(X, Y) \geq 0$ that is:

$$H(X) - H(X|Y) \geq 0$$

that gives the thesis. □

Let $H(X)$ be the uncertainty of X, and $H(X|Y)$ be the conditional uncertainty of X given Y. Mutual information results to be equal to the decrease of the uncertainty of $H(X)$ produced by $H(X|Y)$:

Proposition 3.6

$$I(X, Y) = H(X) - H(X|Y).$$

Proof

$$I(X, Y) = \sum_{x,y} p(x, y) \log \frac{p(x, y)}{p(x)p(y)}$$

but $p(x|y) = \frac{p(x,y)}{p(y)}$, then:

$$I(X, Y) = \sum_{x,y} p(x, y) \log \frac{p(x|y)}{p(x)}$$

from which the following equations hold:

$$\begin{aligned}
I(X, Y) &= \sum_{x,y} p(x, y) \log p(x|y) - \sum_{x,y} p(x, y) \log p(x) \\
&= -H(X|Y) - \sum_{x \in A} \left(\textstyle\sum_{y \in B} p(x, y) \right) \log p(x) \\
&= -H(X|Y) - \sum_{x \in A} p(x) \log p(x) \\
&= -H(X|Y) + H(X) \\
&= H(X) - H(X|Y)
\end{aligned}$$

□

According to the previous proposition, mutual information $I(X, Y)$ can be seen as the uncertainty of X reduced by the uncertainty of X when Y is known. The term $H(X|Y)$ is called **equivocation**, and corresponds to the uncertainty of X given Y.

Mutual information allows us to give a mathematical formulation of the communication process between a sender and a receiver connected by means of a channel of communication where noise is acting disturbing the communication. In this picture sender and receiver are two information sources, or random variables X, Y, respectively, and a quantity, called **channel capacity** C measures the maximum amount of information that can be transmitted through the channel in the time unity, where the maximum is considered in the class of the probability distributions p associated to the sender X:

$$C = \max_{p} I(X, Y)$$

The second Shannon theorem will show that C results to be equivalent to the maximum **transmission rate**, that is, the maximum quantity of information transmitted by the sender, in the time unity, that can be correctly received by the receiver.

3.4 Codes

A **code** is a function from a finite set C of **codewords** or **encodings**, which are strings over a finite alphabet, to a set D of data. In the following, we often will identify a code with the set C of its codewords, by using the term **encoding** for referring to the function from C to D. When encoding is not injective, the code is said to be **redundant**, because different codewords can encode the same datum. When encoding is not surjective, the code is said to be **degenerate**. When encoding is not definite for some codewords, the code is said to be **partial**. In the following discussion, we will often assume 1–1 codes, with 1-to-1 encoding functions.

The **genetic code** is a code that is redundant and partial. It encodes 20 amino acids with terns of symbols, called *codons*, over the alphabet U, C, A, G. Many codons may encode the same amino acid, and there are 3 of the 64 codons (UGA. UAA, UAG) that do not encode any amino acid and are stop signals in the translation from sequences of codons to sequences of amino acids (Table 3.1).

The Standard ASCII has only 7 bits (Table 3.2). Extended ASCII (ISO-Latin-8859) has 8 bits.

In the ASCII code, a character is encoded by summing the column value with the row value. For example: $N \rightarrow 40 + E = 4E$. Then hexadecimal encodings are transformed into binary values.

The **digital information** of datum d, with respect to a code C, is the length of the string $\alpha \in C$ encoding d (the arithmetic mean of the codewords encoding d, if they are more than 1). Digital information is not a good measure of the information of data. In fact, as we will see, given a code, another code can exist encoding the same data with shorter codewords. This means that a measure of information, independent from codes, should require the search for codewords of minimum lengths. One of the main results of Shannon's probabilistic approach is the first Shannon theorem, which explains as the entropy of an information source provides a minimal bound to any digital measure of information.

A code C is **prefix free**, or **instantaneous**, if no codeword of C is a prefix of another codeword of C. When a code is prefix-free, it can be represented by a rooted tree, called **encoding tree**, where codewords are the leaves of the tree. Starting from the root, a number of edges equal to the symbols of the alphabet spring from the root,

Table 3.1 The amino-acids and the corresponding codons

Amino-acid	Name	Letter	Codons
Arg	Arginine	R	$\{CGU, CGC, CGA, CGG, AGA, AGG\}$
Leu	Leucine	L	$\{UUA, UUG, CUU, CUC, CUA, CUG\}$
Ser	Serine	S	$\{UCC, UCU, UCA, UCG, AGU, AGC\}$
Ala	Alanine	A	$\{GCU, GCC, GCA, GCG\}$
Gly	Glycine	G	$\{GGU, GGC, GGA, GGG\}$
Pro	Proline	P	$\{CCU, CCC, CCA, CCG\}$
Thr	Threonine	T	$\{ACU, ACC, ACA, ACG\}$
Val	Valine	V	$\{GUU, GUC, GUA, GUG\}$
Ile	Isoleucine	I	$\{AUU, AUC, AUA\}$
Asn	Aspargine	N	$\{AAU, AAC\}$
Asp	Aspartate	D	$\{GAU, GAC\}$
Cys	Cysteine	C	$\{UGU, UGC\}$
His	Hystidine	H	$\{CAU, CAC\}$
Gln	Glutamine	Q	$\{GAA, GAG\}$
Glu	Glutamate	E	$\{CAA, CAG\}$
Lys	Lysine	K	$\{AAA, AAG\}$
Phe	Phenilananine	F	$\{UUU, UUC\}$
Tyr	Tyrosine	Y	$\{UAU, UAC\}$
Met	Methionine	M	$\{AUG\}$
Trp	Tryptophan	W	$\{UGG\}$

Table 3.2 The Table of ASCII standard code

	0	1	2	3	4	5	6	7	8	9	A	B	C	D	E	F
00	NULL	SOH	STX	ETX	EOT	ENQ	ACK	BEL	BS	TAB	LF	VT	FF	CR	SO	SI
10	DLE	DC1	DC2	DC3	DC4	NAK	SYN	ETB	CAN	EM	SUB	ESC	FS	GS	RS	US
20	SPACE	!	¨	#	$	%	&	’	(	)	*	+	,	–	.	/
30	0	1	2	3	4	5	6	7	8	9	:	;	<	=	>	?
40	@	A	B	C	D	E	F	G	H	I	J	K	L	M	N	O
50	P	Q	R	S	T	U	V	W	X	Y	Z	[	\	]	ê	_
60	‘	a	b	c	d	e	f	g	h	i	j	k	l	m	n	o
70	p	q	r	s	t	u	v	w	x	y	z	{	|	}	˜	DEL

each edge labelled by a symbol of the alphabet of C. The same situation is repeated for all the nodes of the tree that are not leaves. In such a way, each path from the root to a leaf identifies the sequence of the edges of the path, and the sequence of labels is a codeword of C. Therefore, by construction, codewords cannot be substrings of other codewords. The encoding tree of Fig. 3.4 shows that that, passing from a node to their sons, a partition of data is realized, and a leaf is obtained when a single datum is reached. If the partitions are uniformly distributed, the tree reaches a minimal depth, which results to be logarithmic with respect to the size of the set of data.

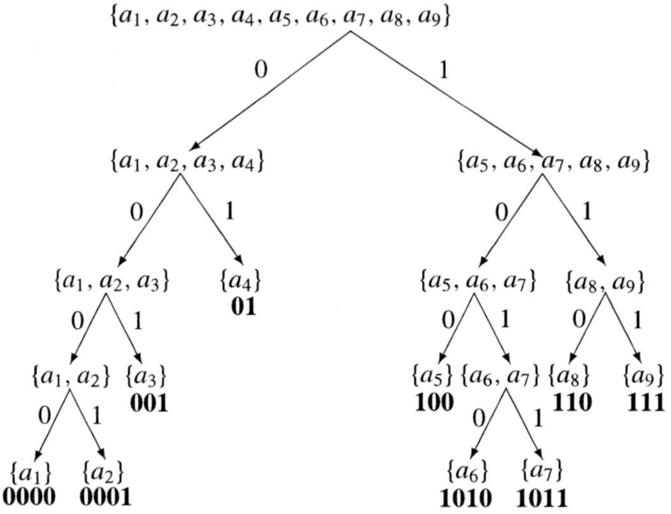

An encoding tree

The above encoding tree shows a very general and universal aspect of strings, which we emphasize with the following proposition.

Proposition 3.7 (Linearization of data) *Any finite set D of data can be represented by a finite sets of strings, called codewords of the elements of D.*

McMillan and Kraft Theorems

A code C is **univocal** if any concatenation of codewords of C gives, in a unique way, the sequence of codewords of that concatenation. This means that do not exist two different sequences of codewords of C that when are concatenated provide the same string over the alphabet of C.

A 1-1 code C, for a set of data D, where α_d is the codeword of d, is **optimal**, with respect to an information source (D, p), if no other code for D exists having a smaller average length L_C, with respect to the probability distribution p:

$$L_C = \sum_{d \in D} |\alpha_d| p(d).$$

The *Kraft norm* $||C||$ of a code C is defined by (k is the cardinality of the alphabet of C, and $|\alpha|$ is the length of α):

$$||C|| = \sum_{\alpha \in C} k^{-|\alpha|}. \tag{3.9}$$

Proposition 3.8 (Kraft inverse theorem) *If $\|C\| \leq 1$ then $\exists\, C'$ such that C' is prefix-free and $\|C'\| = \|C\|$ (the number of encodings of a given length is the same in the two codes).*

Proof Let C be a binary code with the following encoding lengths: $\ell_1, \ell_2, \cdots, \ell_m$. and let $\|C\| \leq 1$ and $\ell_1 < \ell_2 < \ldots < \ell_m$.

First, we observe that, by definition of Kraft norm, if two codes C, C' have the same encoding lengths and the same number of encodings for each length, then $||C'|| = ||C||$. Let us construct C' by composing trees. The first tree T_1 has depth ℓ_1 and 2^{ℓ_1} leaves. Let $\alpha_1 \ldots \alpha_n$ be the encodings of C of length ℓ_1, of course $n \leq 2^{\ell_1}$. In our hypothesis we have also encodings of lengths $\ell_2, \ldots, \ell_m$. This means that some leaves of T_1 are not associated to encodings. Therefore we can construct the trees T_2 for allocating the encodings of length ℓ_2. Without loss of generality, we assume that the root of T_2 is the leftmost leaf of T_1 following all the leaves encoding the codewords of length ℓ_1. Then, allocate in the first leaves of T_2 all the codewords of length ℓ_2. The third tree is added with the same principle, and analogously we can continue until the codewords of all lengths are encoded. This process can continue, because if, at some stage of the process, no leaf would be available, this should contradict the hypothesis that Kraft norm of C is less than or equal to 1. The tree that we obtain at end is, by construction, a code C' with the same norm of the original univocal code C. Moreover, C' is a prefix-free code, because the codewords of C' are associated to the leaves of an encoding tree. □

Example 3.1 The codes of the following table have the same norm, code C is univocal but it is not prefix-free, because (α_3 is a prefix of α_4), while C' is prefix-free (Table 3.3).

The norm of Kraft is for both codes:

$$\|C\| = \|C'\| = \frac{7}{8} < 1$$

Table 3.4 gives an example of non-univocal code, where Kraft norm is less than 1.

Table 3.3 Code C is not prefix-free while C' prefix-free.

	Code C	Code C′
α_1	10	00
α_2	00	01
α_3	11	10
α_4	110	110

Table 3.4 An example of code that is not univocal, but having Kraft norm less than 1

α_1	1
α_2	11
α_3	111
α_4	1111

Proposition 3.9 (McMillan Theorem)

$$If\ C\ is\ univocal\ then\ \|C\| \leq 1$$

Let us introduce some concepts in order to develop the proof.
The concatenation of two codes L_1, L_2 is defined by:

$$L_1 \cdot L_2 = \{\alpha\beta | \alpha \in L_1, \beta \in L_2\}$$

The code C^j with $j > 1$, is the j-power of C obtained by concatenating j copies of C:

$$C^j = \underbrace{C \cdot C \cdot C \cdot \ldots \cdot C}$$

Lemma 3.1 *If C is univocal then* $\left\|C^j\right\| = \|C\|^j$.

Proof In fact:

$$\begin{aligned}
\|C\|^j &= \left(\sum_{i=1}^{n} 2^{-|\alpha_i|}\right)^j \\
&= \prod_{k=1}^{j}\left(\sum_{i=1}^{n} 2^{-|\alpha_i|}\right) \\
&= \sum_{i_1=1}^{n} \cdot \sum_{i_2=1}^{n} \cdot \ldots \cdot \sum_{i_j=1}^{n} 2^{-(|\alpha_{i_1}|+|\alpha_{i_2}|+\cdots+|\alpha_{i_j}|)} \\
&= \sum_{i_1 \cdots i_j \in \{1 \cdots n\}} 2^{-(|\alpha_{i_1} \cdot \alpha_{i_2} \cdots \alpha_{i_j}|)} \\
&= \left\|\mathbf{C^j}\right\|
\end{aligned}$$

□

because when C is univocal:

$$|\alpha_{i_1}| + |\alpha_{i_2}| + \cdots + |\alpha_{i_j}| = |\alpha_{i_1} \cdot \alpha_{i_2} \cdots \alpha_{i_j}|$$

by the uniqueness of the sequence of codewords of C with a given concatenation.

Let max and min be the maximum and minimum of the length of codewords of C, then we have that the maximum number of different lengths of codewords of C^j is $\leq \mathbf{j(max - min) + 1}$.

Proof (*Proof of the theorem of McMillan* [3]) Let C be a univocal binary code with a maximum length of codewords denoted by *max* and minimum length *min* (in the case of a code over an alphabet with more than 2 symbols the proof is analogous). Let C^j, for any $j > 1$, a power of C, then C^j has a number of different lengths of codewords $\ell_1 \dots \ell_m$ that is at most given by $j(max - min) + 1$. In the Kraft norm codewords of a given length give a contribution of less than 1. In fact if in C there are n_1 codewords long l_1, n_2 long l_2, up to n_m long l_m, the norm (in base 2) is:

$$n_1 2^{-l_1} + n_2 2^{-l_2} + \dots + n_m 2^{-l_m}$$

but $n_1 \leq 2^{-l_1}, n_2 \leq 2^{-l_2}, \dots, n_m \leq 2^{-l_m}$, whence

$$n_1 2^{-l_1} + n_2 2^{-l_2} + \dots + n_m 2^{-l_m} \leq 2^{l_1} 2^{-l_1} + 2^{l_2} 2^{-l_2} + \dots + 2^{l_m} 2^{-l_m} \leq m$$

Therefore, the norm of C is at most equal to the maximum number of its lengths:

$$\left\| C^j \right\| \leq m \leq j(max - min) + 1$$

If C is univocal, then

$$\left\| C^j \right\| = \| C \|^j$$

whence

$$\| C \|^j \leq j(max - min) + 1$$

but in this inequality we have on the left j as an exponent, while on the right it is a linear term, therefore it can hold for any $j > 1$ only if:

$$\| C \| \leq 1$$

□

From McMillan's theorem and the inverse Kraft theorem, we get the following corollary.

Corollary 3.1 *If C is univocal, then there exists a prefix-free code C' such that $\|C\| = \|C'\|$.*

3.5 Huffman Encoding

Of course, codes having all codewords of the same length are univocal. However, this requirement is often too strong in many cases. Assume that we want to optimize the average length of strings encoding data emitted by an information source. In this case, it is reasonable to have shorter codewords for data more frequently, allowing longer codewords for data seldom appearing in the source generation.

A code is **optimal**, with respect to an information source, when its probabilistic *average length* $\sum_{\alpha \in C} |\alpha| p(\alpha)$ is minimal. Simple algorithms exist providing codes that are optimal for a given probability distribution.

Assume an information source:

$$(A, p) = \begin{pmatrix} d_1 & \dots & d_n \\ p_1 & \dots & p_n \end{pmatrix}$$

Huffman Encoding is based on the cardinality k of the alphabet (here we assume a binary code).

1. Let us order the data according to the increasing order of their probabilities.

$$p_1 \leq p_2 \leq \cdots \leq p_{n-1} \leq p_n$$

2. Replace the two data with the lowest probabilities with only one datum having the sum of their probabilities.

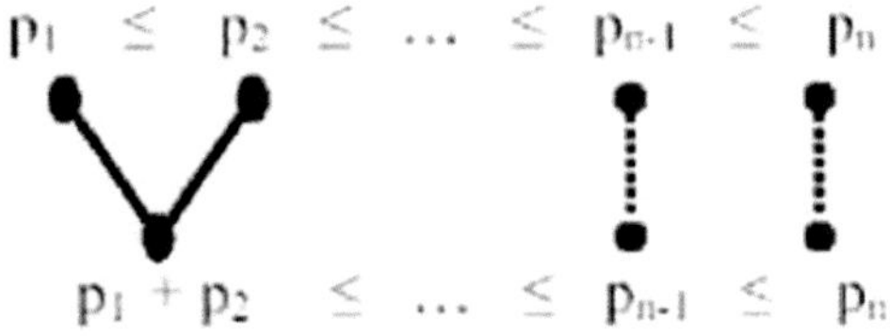

3. An edge is labeled by 0 and the other with 1. Go to step 1.
4. Stop when only one datum is available. The encodings are the sequences of labels of each path from the leaves to the root.

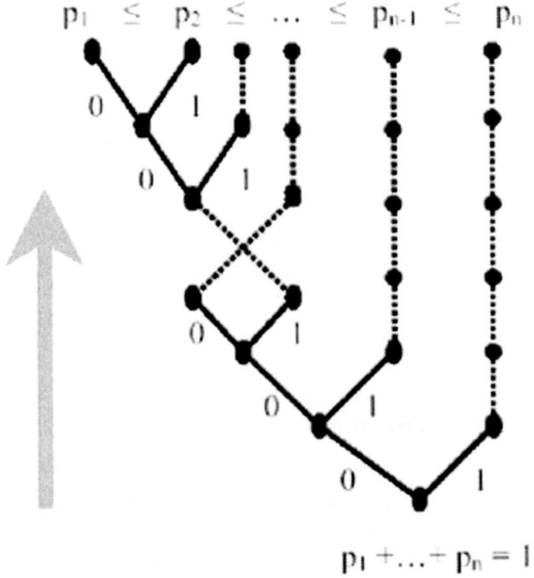

Example 3.2 Consider the probabilities given in the following table, an Huffman code resulting from these probabilities is given in the following figure.

	Probability
a	0.46
b	0.18
c	0.09
d	0.09
r	0.18

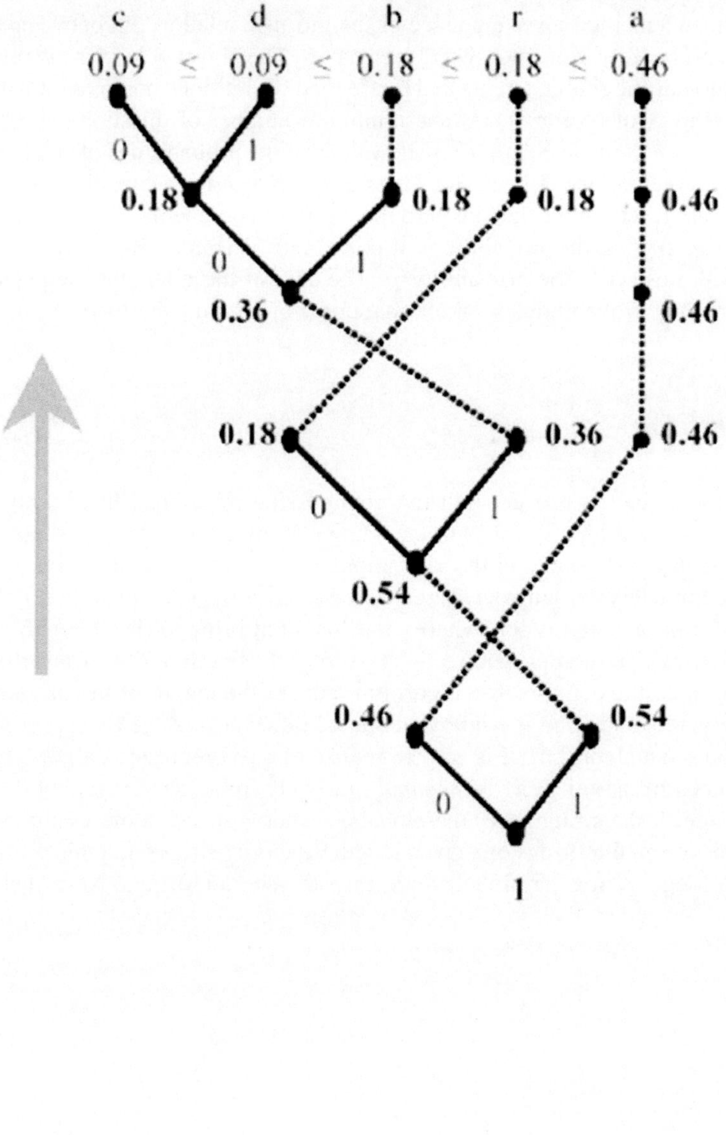

	Code C'
a	0
b	111
c	1100
d	1101
r	10

The Huffman method is not deterministic (probabilities can be aggregated, and edge labels can be used, in different ways). However, It can be shown that Huffman codes are optimal codes [3].

Huffman's method gives evidence of the intrinsic relationship between information and probability. Consider two players A, B. The player A knows the probability distribution of the source (X, p) and the second player does not know it but can ask questions to A of type $x \leq a$?. The minimum number of questions necessary, on average, to guess the most approximate values of the probabilities of p has $H(X, p)$ as a lower bound. If we organize the questions of B to A in an encoding tree. It can be shown that the lengths of the tree with the minimum number of questions necessary, on average, to guess the probabilities of p is a Huffman binary encoding of the values of X, with respect to the probability p. The sum of these lengths weighted by the corresponding probabilities is an optimal upper approximation to $H(X, p)$.

Compression

Optimal encoding is related with the compression of strings. In general terms, a compression algorithm is a way of encoding a string α by a string β shorten than α (let us assume β to belong to the same alphabet of α) such that the original string α can be completely and univocally recovered from β by a *decompression* algorithm (partial forms of compression, where some loss of information is tolerated, are useful in some cases). The compression ratio is given by the fraction $|\beta|/|\alpha|$, and the smaller it is, the more the compression algorithm reduces the length of the original string. Of course, this reduction is paid by a computational cost of the compression and/or decompression algorithm. It is easy to realize that a compression algorithm cannot be universal because it can give a small ratio only for some strings, but this cannot happen for all the strings over the alphabet, otherwise, the compression algorithm could encode all the strings of a given length by shorter strings, but this is impossible because longer strings are much more than the shorter strings. Nevertheless, it is often interesting that compression could really compress some classes of strings for which a more compact way is required. Methods of compression are usually based on three different principles (in many aspects related). The first is based on a known distribution of the probabilities of some words occurring in the string. In this case, it is enough to define short encodings for words with high occurrence probabilities and longer encodings for those with low probabilities. The second method is based on dictionaries. If we know a dictionary of words occurring in a string, then we memorize this dictionary and replace the words with the corresponding encodings within the text. This provides a compression when we have, for example, one thousand binary words of length 20. In fact, one thousand objects need only binary strings of length 10, therefore we get a reduction of 10 bits for each word occurrence. The third method is based on a rearrangement of a string from which the original string can uniquely be recovered, in such a way that in the rearranged string similar substrings are contiguous. If this happens, an element of type α^n can be encoded by encoding α and n, that need $|\alpha|$ symbols plus $O(\lg n)$ symbols instead than $n|\alpha|$ symbols (the n copies of α scattered in the original sequence).

3.6 First Shannon Theorem

Now we show that the notion of entropy allows us to determine a lower bound to the compression. Let us give a preliminary lemma.

Lemma 3.2 (Logarithm Lemma)

Let $p_1, \ldots, p_m$ and $q_1, \ldots, q_m$ be two probability distributions defined on the same set of m *data:*

$$\sum_{i=1}^{m} p_i = \sum_{i=1}^{m} q_i = 1.$$

For any base k of the logarithms we have:

$$-\sum_{i=1}^{m} p_i \log_k p_i \leq -\sum_{i=1}^{m} p_i \log_k q_i$$

Proof As it is shown in the following diagram the logarithm is always under the line $y = x - 1$, therefore:

$$\ln x \leq x - 1 \quad \forall x \in R$$

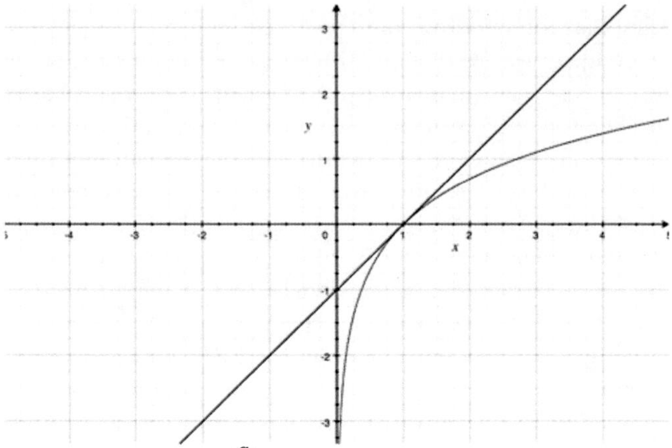

If we consider the value $x = \frac{q_i}{p_i}$, then we obtain:

$$\ln \frac{q_i}{p_i} \le \frac{q_i}{p_i} - 1$$

and multiplying both members by p_i, and summing we have:

$$\sum_i p_i \ln \frac{q_i}{p_i} \le \sum_i p_i \left(\frac{q_i}{p_i} - 1 \right)$$

But, according to the hypothesis: $\sum_{i=1}^{m} p_i = \sum_{i=1}^{m} q_i = 1$, it holds:

$$\sum_i p_i \ln \frac{q_i}{p_i} \le \overbrace{\sum_i q_i}^{1} - \overbrace{\sum_i p_i}^{1}$$

$$\sum_i p_i \ln \frac{q_i}{p_i} \le 0$$

whence:

$$\sum_i p_i \log_k \frac{q_i}{p_i} \le 0$$

(we can pass to any other base $k' > 1$ by multiplying for the constant $\lg_{k'} k$) by the logarithm properties:

$$- \sum_i p_i \log_k p_i + \sum_i p_i \log_k q_i \le 0$$

hence:

$$-\sum_i p_i \log_k p_i \leq -\sum_i p_i \log_k q_i$$

□

Given a code C encoding data of an information source, the probability distribution of data is naturally associated with the codewords of C. We define L_C as the **mean length** of C by:

$$L_C = \sum_{\alpha \in C} |\alpha| p(\alpha).$$

Of course, the shorter is L_C, the most compressed is the encoding of C. The following theorem establishes a lower bound to the mean length L_C.

Theorem 3.1 (First Shannon Theorem)
Given an information source (A, p) and a prefix-free code C over A, the following inequality holds:

$$L_C \geq H(A, p)$$

Proof Let us consider the following distribution q (where a ranges in the set of data):

$$q(a) = \frac{k^{-|c(a)|}}{\|C\|}$$

let also $c(a)$ be the encoding of a and $\|C\|$ the Kraft norm of C. From the definition of $\|C\|$ it follows that q is a probability distribution.

If k is the cardinality of the alphabet A, we have (the inequality follows from the logarithm lemma):

$$\begin{aligned}
H(A, p) &= -\sum_{a \in A} p(a) \log_k p(a) \\
&\leq -\sum_{a \in A} p(a) \log_k q(a) \\
&= -\sum_{a \in A} p(a) \log_k \frac{k^{-|c(a)|}}{\|C\|} \\
&= -\sum_{a \in A} p(a)(\log_k k^{-|c(a)|} - \log_k \|C\|) \\
&= -\sum_{a \in A} p(a) \log_k k^{-|c(a)|} + \sum_{a \in A} p(a) \log_k \|C\| \\
&= \sum_{a \in A} p(a)\, |c(a)| + \log_k \|C\| \left(\sum_{a \in A} p(a) \right).
\end{aligned}$$

Being $\sum_{a \in A} p(a) = 1$, we obtain:

$$H(A, p) \leq L_C + \log_k \|C\|$$

But C is prefix-free, so $\|C\| \leq 1$, whence $\log_k \|C\| \leq 0$, that is, the thesis follows. □

3.7 Typical Sequences

Given an information source (A, p), we say **typical** a sequence where the frequency of every datum coincides with the probability of the information source.

3.7.1 AEP (Asymptotic Equipartition Property) Theorem

Let us consider an information source (A, p) emitting data at steps: $1, 2, \ldots n, \ldots$. We ask what is the probability of a typical sequence of length n emitted by the source (A, p). We will show that this probability is directly related to the entropy of the source. The following proposition holds.

Proposition 3.10 *1. The probability of each typical sequence of length N is* 2^{-NH}*;*

2. *If N tends to the infinite, then the probability of a sequence long N approaches the probability of a typical sequence with the same length;*
3. *For N tending to the infinite the cardinality of typical sequences approaches* 2^{NH}.

Proof Let us consider sequences of length N over data $A = \{a_1 \ldots a_m\}$. Let N_i be the number of occurrences a_i.

First point.
In any typical sequence α, for every $i \in \{1, \ldots, m\}$, $N_i = Np_i$, so, its probability is:

$$p(\alpha) = p_1^{Np_1} \cdot p_2^{Np_2} \cdot \ldots \cdot p_m^{Np_m}.$$

Therefore (for a binary base) we have:

$$\begin{aligned}
\log p(\alpha) &= \log(p_1^{Np_1} \cdot p_2^{Np_2} \cdot \ldots \cdot p_m^{Np_m}) \\
&= \log p_1^{Np_1} + \log p_2^{Np_2} + \cdots + \log p_m^{Np_m} \\
&= \sum_{i=1}^{m} \log p_i^{Np_i} \\
&= \sum_{i=1}^{m} Np_i \log p_i \\
&= N \cdot \sum_{i=1}^{m} p_i \log p_i \\
&= N(-H) \\
&= -NH
\end{aligned}$$

whence

$$p(\alpha) = 2^{-NH}.$$

Second point.
From the law of large numbers with the increasing of N we have with probability tending to one:

$$\frac{N_i}{N} \to p_i \quad \Rightarrow \quad N_i \to p_i N$$

therefore, increasing N any sequence approaches to a typical sequence of the same length.

Third point.
The probability of any typical sequence of length N is 2^{-NH}, therefore from point (2) according to which for $N \to \infty$ any sequence approaches a typical sequence, the number of typical sequences is the inverse 2^{-NH}, that is, 2^{NH}. □

Example 3.3 Let us consider binary sequences where:

$$p = \{p_0, p_1\}, \, p_0 = 0.89, \, p_1 = 0.11$$

where p_0 e p_1 are the probabilities of 0 and 1 respectively.

Let us consider sequences of length 100.

- In 100 bits ⇒ we can assume 89 times 0 and 11 times 1;
- In 200 bits ⇒ we can assume 178 times 0 and 22 times 1 .

Let us evaluate the sequences that differ at most in two bits from the ratio 89/11. According to the binomial distribution $B(n, p)$ (n bits, p probability of 1):

$$P(X = k) = p_x(k) = \binom{n}{k} * p^k (1-p)^{n-k}$$

$$\begin{array}{lll}
91 - 9 & \Rightarrow & p(\alpha) = \binom{100}{9} * 0.91^{91} * 0.09^{9} = 0.138 \\
90 - 10 & \Rightarrow & p(\alpha) = \binom{100}{10} * 0.90^{90} * 0.10^{10} = 0.132 \\
89 - 11 & \Rightarrow & p(\alpha) = \binom{100}{11} * 0.89^{89} * 0.11^{11} = 0.127 \\
88 - 12 & \Rightarrow & p(\alpha) = \binom{100}{12} * 0.88^{88} * 0.12^{12} = 0.122 \\
87 - 13 & \Rightarrow & p(\alpha) = \binom{100}{13} * 0.87^{87} * 0.13^{13} = 0.118
\end{array}$$

Summing these probabilities we have a value of 0.637. Therefore, with a probability greater than 60% we have sequences that are about 10^{-15} of the total number of possible sequences, which are about 10^{30}.

Typical sequences provide a further characterization of entropy as the limit (with respect to the length) of the ratio between the logarithm of the number of typical sequences and the logarithm of the number of all the possible sequences of a given length (logarithms are with respect to a base equal to the number of symbols of the alphabet).

3.8 Second Shannon Theorem

The second Theorem of Shannon concerns the transmission of information along a channel. Shannon shows that even if the channel is affected by an error, it is possible, under suitable conditions, to transmit information by reaching an error probability that approaches zero, as the redundancy of encoding increases. In other words, a suitable kind of redundancy can compensate for error transmission, by ensuring methods of reliable communication. This theorem is strongly based on the notion of a typical sequence for an information source.

The **capacity** C of a communication channel is the maximum value of mutual information between the random variables associated with the information sources of the sender and of the receiver, when the probability distribution associated with the sender source ranges on the set of all possible probability distributions (this maximum can be proved to exist):

$$C = \max_{P}\{I(X||Y)\}.$$

Channel Encoding

A transmission code is a code used for transmitting, with redundancy that introduces more symbols than those necessary to represent all the data, in such a way that it is possible to contrast the loss of information due to the noise affecting the channel. The Shannon second theorem will explain that in principle this is a possible solution for ensuring an error-free transmission even in presence of noise. The following developments of auto-correcting codes show that this possibility is effectively realizable. This is one of the most exciting validations of the correctness of Shannon's analysis of communication in terms of entropic concepts.

The **transmission error** is the number of times a message is decoded by the receiver with a message different from the message that was sent.

If $\alpha = a_1, a_2 \ldots, a_n$ is sent but $\beta = a_1, b_2, a_3 \ldots, a_n$ with $a_2 \neq b_2$ is received, then the channel introduced an error.

Let's consider a binary code C encoding 10 messages. Of course, the length of each encoding has to be greater than 3. In fact, strings of length 3 can at most encode 8 messages. A transmission code for a number M of messages uses blocks, that is strings, of k symbols of an alphabet A having a length n such that $k^n \geq M$.

A **transmission code** of type (M, n) is a code encoding a number M of possible messages, each of them by a sequence of length n (over a finite alphabet), called also a **block** (all codewords have the same length).

Let us assume to transmit with a code of type (M, n) and that the alphabet A of the code has k symbols, then k^n is the number of possible messages of length n that can be formed over the alphabet A, therefore the inequality $M \leq k^n$ has to hold.

Given a transmission code of type (M, n), its **Transmission Rate** R is given by:

$$R = \frac{log_k M}{n}$$

$$M \leq k^n$$

$$\log_k M \leq \log_k k^n$$

therefore:

$$\log_k M \leq n$$

that is:

$$\frac{\log_k M}{n} \leq 1$$

$$R \leq 1$$

R represents the percentage of the used transmission band. When $R < 1$, then some *redundancy* is introduced. This means that in the case of a binary alphabet, more bits than those necessary for encoding the messages are used.

The rate R is **reachable** by a family of transmission codes $\{(M, n) \mid n \in \mathbb{N}\}$ if a succession of codes of this family exists such that the transmission error approaches to zero as n increases.

When a code $(2^{nR}, n)$ is used, the transmission rate is given by:

$$\frac{\log_2 2^{nR}}{n} = \frac{nR}{n} = R$$

Let us consider a sender information source X, a channel of capacity C, and a transmission process toward a receiver represented by a random variable Y, with a transmission rate R. The following theorem holds.

Theorem 3.2 (Second Shannon Theorem)

Let $R < C$, then there exists a sequence of codes $(2^{nR}, n)$ for which the rate R is reachable.

Proof (*Informal Proof*) In a transmission process, let X be the random variable of the sender and Y that one of the receivers. According to the AEP theorem, we can consider only typical sequences. The average number of *X-typical* sequences, given one *Y-typical* sequence, received along the transmission, is expressed by:

$$2^{nH(X|Y)}.$$

The following picture visualizes this fact, where sequences are identified by points. On the left there are *X-typical* sequences, while on the right there are the corresponding received *Y-typical* sequences.

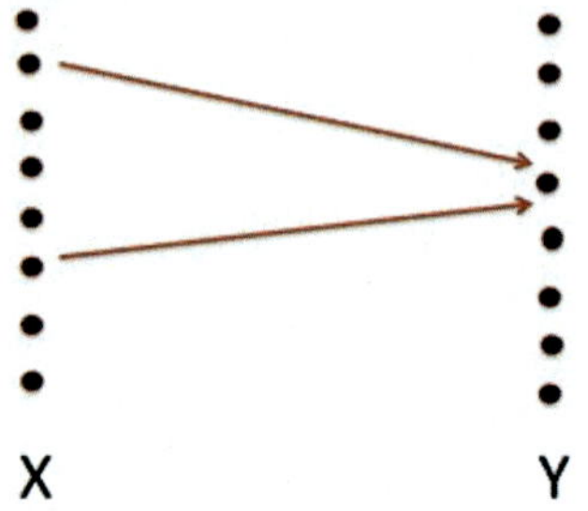

The average probability that one *X-typical* is chosen as the origin of one *Y-typical* is the ratio between: (i) the number of *X-typical* given a *Y-typical* sequences, and (ii) the number *X-typical*. This ratio can be expressed, in terms of conditional entropy, in the following way:

$$\frac{2^{nH(X|Y)}}{2^{nH(X)}} = 2^{nH(X|Y)-H(X)}.$$

From a property of *mutual information* we have:

$$I(X, Y) = H(X) - H(X|Y)$$

therefore, the average probability of determining one *X-typical* sequence as origin of one *Y-typical* is:

$$2^{-nI(X,Y)}$$

From this probability, we can evaluate the probability of error. In fact, the number of wrong messages is equal to the number of possible messages minus 1 (the only right one), that is:

$$2^{nR} - 1 \approx 2^{nR}$$

therefore, the error probability E is given by:

$$E = 2^{nR} * 2^{-nI(X,Y)}$$

$$E = 2^{nR-nC}$$

$$E = 2^{-n(C-R)}$$

whence if $R < C$ the error probability E approaches to zero. □

3.9 Signals and Continuous Distributions

Waves are periodical functions that propagate in space. For this reason, they can transport data and therefore can be used as channels for transmitting the information. However, their nature is continuous. Hence, if we want to apply to waves the entropic concepts, we need to extend them to the case of continuous distributions. This can be done in a very natural manner, by replacing sums with integrals. Namely, if $f(x)$ is a continuous distribution defined over a set A of values, then its entropy $H(f)$ is given by:

$$H(f) = \int_A f(x) \lg f(x) dx.$$

Analogously, if f, g are distributions over the same set of values A, then the entropic divergence of f with respect to g is given by:

$$D(f, g) = \int_A f(x)[\lg f(x) - \lg g(x)]dx.$$

and the mutual information of random variable X with respect to a random variable Y, on the same range of values, is given by:

$$I(f, g) = \int_A f(x)[\lg f_{X,Y}(x, y) - \lg(f_X(x) f_Y(y))]dx.$$

where $f_{X,Y}(x, y)$ is the distribution of the joint events, while f_X and f_Y are the (marginal) probabilities of the same events.

Let $N(x)$ denote the normal probability distribution $N(x) = \frac{1}{\sqrt{2\pi}\sigma} e^{-\frac{x^2}{2\sigma^2}}$, and $H(f)$ be the Shannon entropy of a probability distribution f (here $D(f, N)$ is the continuous version of Kullback-Leibler divergence). The following propositions are proven in [3].

Proposition 3.11

$$H(N) = \frac{1}{2} \ln(2\pi e \sigma^2) \tag{3.10}$$

Proof ([3]).

$$H(N) = -\int_{-\infty}^{+\infty} N(x) \ln N(x) dx$$

$$= \int_{-\infty}^{+\infty} - N(x) \ln \frac{e^{-\frac{x^2}{2\sigma^2}}}{\sqrt{2\pi\sigma^2}} dx$$

$$= \int_{-\infty}^{+\infty} - N(x)[-\frac{x^2}{2\sigma^2} - \ln \sqrt{2\pi\sigma^2}]dx$$

$$= \int_{-\infty}^{+\infty} \frac{N(x)x^2}{2\sigma^2} dx + \ln \sqrt{2\pi\sigma^2} \int_{-\infty}^{+\infty} N(x)dx$$

$$= \frac{E(x^2)}{2\sigma^2} + \ln \sqrt{2\pi\sigma^2} \cdot 1$$

$$= \frac{1}{2} + \frac{1}{2} \ln 2\pi\sigma^2$$

$$= \frac{1}{2} \ln 2\pi e\sigma^2.$$ □

Proposition 3.12 *Normal distributions are those for which entropy reaches the maximum value in the class of probability distributions having the same variance.*

Proof ([3]) Let f denote any probability distribution of variance $Var(f) = \sigma^2$.

$$D(f, N) = \int_{-\infty}^{+\infty} f(x) \ln \frac{f(x)}{N(x)} dx$$

$$D(f, N) = \int_{-\infty}^{+\infty} f(x) \ln f(x)dx - \int_{-\infty}^{+\infty} f(x) \ln N(x)dx$$

$$D(f, N) = \int_{-\infty}^{+\infty} f(x) \ln f(x)dx - \int_{-\infty}^{+\infty} f(x) \ln \frac{e^{-\frac{x^2}{2\sigma^2}}}{\sqrt{2\pi\sigma^2}} dx$$

$$D(f, N) = -S(f) - \int_{-\infty}^{+\infty} f(x) \ln e^{-\frac{x^2}{2\sigma^2}} dx + \ln \sqrt{2\pi\sigma^2} \int_{-\infty}^{+\infty} f(x)dx$$

$$D(f, N) = -S(f) + \frac{1}{2\sigma^2} \int_{-\infty}^{+\infty} f(x)x^2 dx + \frac{1}{2} \ln 2\pi\sigma^2 \cdot 1$$

$$D(f, N) = -S(f) + \frac{1}{2\sigma^2} Var(f) + \frac{1}{2} \ln(2\pi\sigma^2) \; (Var(f) \leq \sigma^2)$$

$$D(f, N) \leq -S(f) + \frac{1}{2} + \frac{1}{2} \ln(2\pi\sigma^2)$$

$$D(f, N) = -S(f) + \frac{1}{2}(\ln e + \ln(2\pi\sigma^2))$$

$$D(f \| N) \leq -S(f) + \frac{1}{2} \ln(2\pi e\sigma^2) \; (D(f \| N) \geq 0)$$

therefore $H(f) \leq H(N)$. □

The notion of a channel is very broad, it is any medium through which data, which are always physical objects, move in the space from one place where the sender is located to another place where the receiver is located. If data are sounds they propagate in the air, if are molecules they may float in the water. The most interesting case in many applications is propagation through electromagnetic waves.

These waves propagate in space without no other medium. In this way, data can be transmitted by modulating waves by altering their forms, in such a way that the alterations correspond, according to specific criteria, to "write" on them codewords. When the signal is received by the receiver, the alterations are "read" and codewords are recovered by obtaining the data that were transmitted by means of the signals. From a mathematical point of view, a signal is a periodic, in time, function that propagates in space. The **signal band** is the quantity of information (number of bits) that the signal can transmit in the unit of time. In this section, we will show that this quantity is related to the frequency of the signal (sampling theorem).

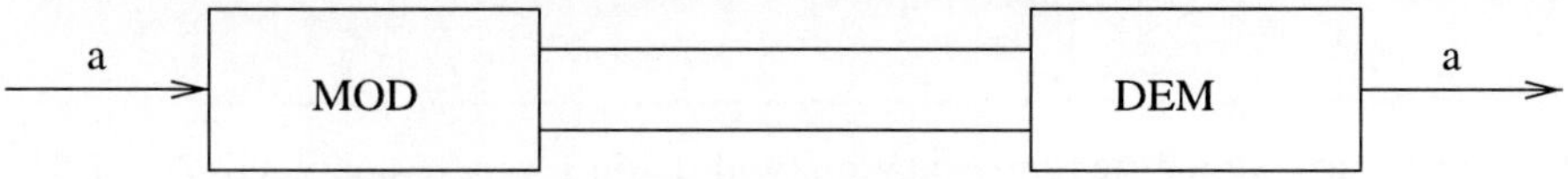

This is the general schema of a communication:

Source	data-encoding
Sender	message-encoding
Emittent	channel-encoding
Signal	modulation
Channel	transmission
Recipient	demodulation
Receiver	message-decoding
Destination	data-decoding

3.9.1 *Fourier Series*

A periodic continuous (or piecewise continuous) function $f(t)$ $(f(t) = f(t + n * T))$, can be completely represented as an infinite sum of sine and cosine functions.

$$f(t) = \underset{constant}{a_0} + \underbrace{a_1 \sin(\omega t) + b_1 \cos(\omega t)}_{fundamental\ harmonic} + \underbrace{\sum_{n=2}^{\infty} a_n \sin(n\omega t) + \sum_{n=2}^{\infty} b_n \cos(n\omega t)}_{other\ harmonics}$$

where:

- T is the period and $\nu = \frac{1}{T}$ is the frequency (number of periods in a second)
- $\omega = 2\pi\nu$ is called *pulsation*
- the coefficients of Fourier series are given by:

$$a_0 = \frac{1}{T}\int_0^T f(t)dt$$
$$a_n = \frac{2}{T}\int_0^T f(t)\sin(n\omega t)dt$$
$$b_n = \frac{2}{T}\int_0^T f(t)\cos(n\omega t)dt$$

For example, let $f(t)$ be given by:

$$f(t) = \underbrace{4 * \sin(\frac{2\pi}{50}t) + 6 * \cos(\frac{2\pi}{50}t)} + \underbrace{3 * \sin(2 * \frac{2\pi}{20}t) + 8 * \cos(2 * \frac{2\pi}{20}t)}$$

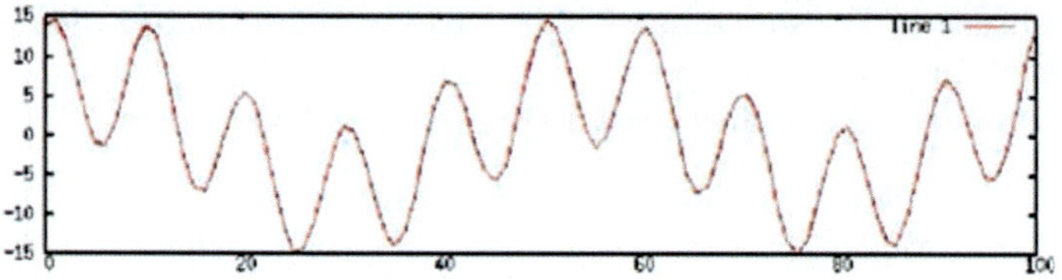

In a Fourier trigonometric representation (with 1, $\sin x$, $\cos x$, $\sin 2x$, $\cos 2x$...). The sum converges to the values of f and, in the case of discontinuity points, the sum converges to the average of left and right limits.

The signal above can be composed by the following components.

$$g(t) = \underbrace{4 * \sin(\frac{2\pi}{50}t) + 6 * \cos(\frac{2\pi}{50}t)}_{\text{fundamental harmonic}}$$

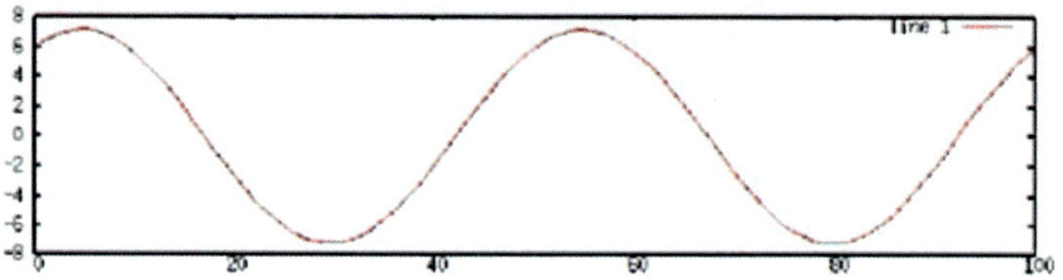

$$h(t) = \underbrace{3 * \sin(2 * \frac{2\pi}{20}t) + 8 * \cos(2 * \frac{2\pi}{20}t)}_{\text{second harmonic}}$$

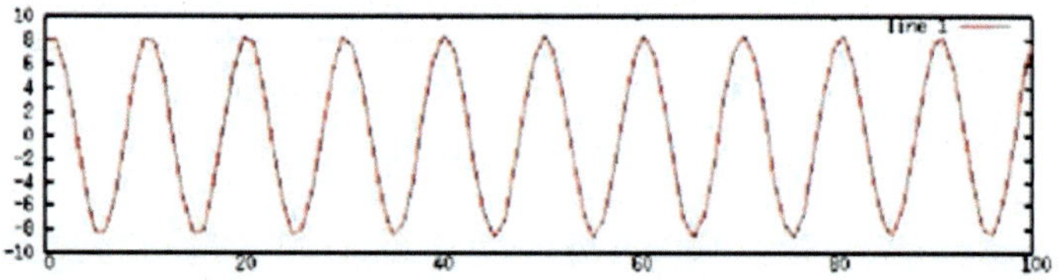

The two harmonics are:

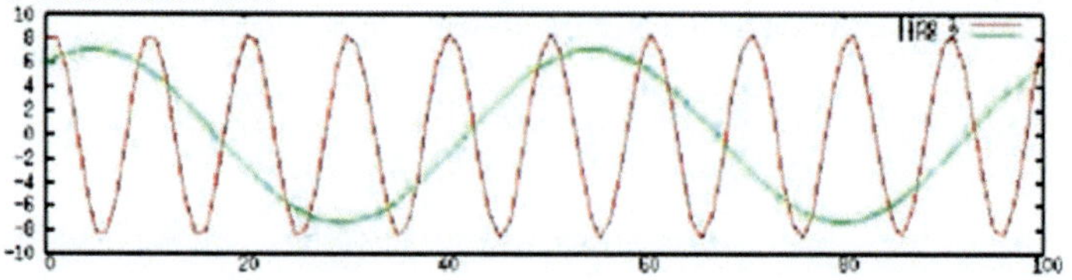

summing them in all the points we obtain the composite signal.

Let us recall Euler formulae for complex numbers, where ϑ is real and i is the imaginary unit, are:

$$e^{i\vartheta} = \cos\vartheta + i\sin\vartheta$$

$$e^{-i\vartheta} = \cos\vartheta - i\sin\vartheta$$

$$\sin\vartheta = \frac{e^{i\vartheta} - e^{-i\vartheta}}{2i}$$

$$\cos\vartheta = \frac{e^{i\vartheta} + e^{-i\vartheta}}{2}$$

Fourier summation in the field of complex numbers, by using Euler formulae, gives:

$$f(t) = \sum_{n=-\infty}^{+\infty} c_n e^{in\omega t}$$

$$c_n = \frac{1}{2T}\int_{-T}^{T} f(t) e^{-in\omega t} dt$$

Now we explain as Fourier representations are naturally definable in the general setting of Hilbert spaces. A vector space over an algebraic field is a structure that axiomatically expresses the nature of $\mathbf{R}^n$ (n-tuples of real numbers) with the sum (of each component), a product by a scalar value (multiplying all the components for the same value) and the zero vector with all components zero. A basis for such a space is a set of vectors such that any other vector can be obtained by sum and product by scalars from some vectors of the basis. A **Hilbert space** is a vector space that is complete, that is, any Cauchy sequence of vectors converges to some vector of the space, and moreover, a scalar product is defined, that associates to any pair of vectors u, v a real value $< u|v >$. A special case of Hilbert space is the functions $L^2[a, b]$ defined on real numbers such that their square is integrable in some closed interval $[a, b]$. This space is extended to the complex functions with the scalar product, where f^* is the complex conjugate of the function f:

$$< f|g >= \int_{a}^{b} f(x) f^*(x)$$

A basis of vectors is orthonormal if the scalar product is null between any two vectors of the basis, and moreover $< v|v >= 1$ for any v of the basis.

In the case of a finite basis, we have:

$$v = \sum_{n=-1}^{N} c_n e_n$$

and for every v in the basis and $n = 1, \ldots N$:

$$c_n =< v|e_n >$$

In any Hilbert space, the following propositions hold.

Proposition 3.13 *Given a orthonormal basis of a Hilbert space, for every vector v of a basis B we have:*

$$v = \sum_{e \in B} < v|e > e$$

Proposition 3.14 *In the space $L^2[l, -l]$ of complex function square-integrable complex functions the following functions are a basis for that space:*

$$\frac{1}{\sqrt{2l}}\{e^{i\frac{\pi}{l}nx} \mid n \in \mathbf{N}\}.$$

Proof Let $n \neq m$ and $n - m = k$.

$$< f|g >= \frac{1}{2l}\int_{-l}^{l} e^{-i\frac{\pi}{l}nt} e^{i\frac{\pi}{l}nt} dt = \frac{1}{2l}\int_{-l}^{l} dt = 1$$

$< f|f >= \frac{1}{2l}\int_{-l}^{l} e^{-i\frac{\pi}{l}nt} e^{i\frac{\pi}{l}mt} dt =$
$\frac{1}{2l}\int_{-l}^{l} e^{i\frac{\pi}{l}kt} dt =$
$\frac{l}{2ilk\pi}\left[e^{i\frac{k\pi t}{l}}\right]_{-l}^{l} =$
$\frac{e^{ik\pi} - e^{-ik\pi}}{2ik\pi} = \frac{\sin k\pi}{k\pi} = 0.$ □

From the two propositions above the Fourier's representation follows as a consequence of the Hilbert space structure of circular functions (l instead of T is used in order to be coherent with the following notation).

Proposition 3.15 *Every function in $L^2[-l, l]$ can be expressed as a linear combination (with complex coefficients) of circular functions.*

$$f(x) = \sum_{n=-\infty}^{+\infty} c_n e^{i\frac{\pi}{l}nx} \quad (3.11)$$

$$c_n = \frac{1}{2l}\int_{l}^{l} f(x) e^{-i\frac{\pi}{l}nx} dx \quad (3.12)$$

Proof Using the basis of $L^2[-l, l]$:

$$f(x) = \sum_{e \in B} < f(x)|\frac{1}{\sqrt{2l}}e^{i\frac{\pi}{l}nx} > \frac{1}{\sqrt{2l}}e^{i\frac{\pi}{l}nx}.$$

□

We remark that for technical reasons, which we do not explicit more precisely, in the context of Hilbert space a notion of integral more general than Riemann integral is used, which was introduced by Lebesgue in the context of the general measure theory.

3.10 Fourier Transform

The Fourier Transform is obtained with a process that generalizes the coefficients in the Fourier representation of functions.

Let

$$\omega_n = n\frac{\pi}{l}$$

then:

$$c_n = \frac{1}{2l}\int_{-l}^{l} f(x)e^{-i\omega_n x}dx$$

Let assume that when l goes to the infinite the integral approaches to a finite value, and replace coefficients c_n by $c_n\frac{l}{\pi}$, by obtaining:

$$c_n\frac{l}{\pi} = \frac{1}{2\pi}\int_{-l}^{l} f(x)e^{-i\omega_n x}dx$$

now replacing c_n by $c(\omega)$ we obtain:

$$c(\omega) = \frac{1}{2\pi}\int_{-\infty}^{\infty} f(x)e^{-i\omega x}dx$$

$$f(x) = \sum_{n=-\infty}^{+\infty} c_n\frac{l}{\pi}e^{i\omega_n x}\frac{\pi}{l}$$

where replacing $c_n l/\pi$ by $c(\omega_n)$ we have:

$$f(x) = \sum_{n=-\infty}^{+\infty} c(\omega_n) e^{i\omega_n x} \frac{\pi}{l}$$

when in this sum l goes to the infinite we obtain the following integral:

$$f(x) = \int_{-\infty}^{\infty} c(\omega) e^{i\omega x} dx.$$

In the end, we obtain the following pair where the complex value $F(\omega)$ generalizes the coefficients of Fourier representation. The next theorem shows the advantage of such a generalization.

$$\underbrace{\begin{aligned} f(t) &= \int_{-\infty}^{+\infty} F(\omega) e^{i\omega t} d\omega \quad (1) \\ F(\omega) &= \frac{1}{2\pi} \int_{-\infty}^{+\infty} f(t) e^{-i\omega t} dt \quad (2) \end{aligned}}$$

Pair of Fourier Transforms

$F(\omega)$ is the Fourier Transform of $f(t)$) and $f(t)$ is the anti-transform of $F(\omega)$. It can be shown that the pair (f, F) is completely determined by each of the two components.

3.11 Sampling Theorem

The following theorem highlights the role of signals in transporting information.

Proposition 3.16 (Sampling Theorem (Nyquist-Shannon)) Let $f(t)$ *be a* periodic function with integrable square, in its period, and with limited band W (components of frequencies outside $[-W, +W]$ are zero), then $f(t)$ is completely determined by the values assumed at time intervals with distance $\frac{1}{2W}$: $\left[0, \frac{1}{2W}, 2\frac{1}{2W}, 3\frac{1}{2W}, \ldots\right]$.

Proof Using the Fourier transform we have:

$$f(t) = \int_{-\infty}^{+\infty} F(\omega) e^{i\omega t} d\omega$$

If we consider frequencies in the interval $[-W, +W]$, then we integrate between $-a = -2\pi W$ ad $a = +2\pi W$:

$$f(t) = \int_{-a}^{+a} F(\omega)e^{i\omega t} d\omega \tag{3.13}$$

the function $f(t)$ in the points $-\frac{\pi}{a}n$ is:

$$f(-\frac{\pi}{a}n) = \int_{-a}^{+a} F(\omega)e^{-i\omega\frac{\pi}{a}n} d\omega. \tag{3.14}$$

Now let us consider the anti-transform $F(\omega)$ within $[-a, a]$:

$$F(\omega) = \sum_{n=-\infty}^{+\infty} d_n e^{i\frac{\pi\omega}{a}n} \tag{3.15}$$

with coefficients d_n given by Eq. (3.12):

$$d_n = \frac{1}{2a}\int_{-a}^{a} F(\omega)e^{-i\frac{\pi\omega}{a}n} d\omega \tag{3.16}$$

but from Eqs. (3.13) and (3.14) we have:

$$d_n = \frac{1}{2a} f\left(-\frac{\pi}{a}n\right). \tag{3.17}$$

Therefore, the values of f in $-\frac{\pi}{a}n$ coincide with the coefficients of the Fourier Transform of f. But this transform completely determines f. In conclusion, the values of f in $\pi/a = 1/2W$ allow us to recover the whole function f.

If we substitute the right member of (3.17) in (3.15):

$$F(\omega) = \sum_{n=-\infty}^{+\infty} \frac{1}{2a} f(-\frac{\pi}{a}n)e^{i\frac{\pi\omega}{a}n} \tag{3.18}$$

and then the right member of (3.18) in (3.13):

$$f(t) = \int_{-a}^{+a} \left(\frac{1}{2a}\sum_{n=-\infty}^{+\infty} f(\frac{\pi}{a}n)e^{-i\frac{\pi\omega}{a}n}\right) e^{i\omega t} d\omega \tag{3.19}$$

therefore:

$$f(t) = \frac{1}{2a} \sum_{n=-\infty}^{+\infty} f(\frac{\pi}{a}n) \int_{-a}^{+a} e^{i\omega(t-\frac{\pi}{a}n)} d\omega \tag{3.20}$$

giving:

$$f(t) = \frac{1}{2a} \sum_{n=-\infty}^{+\infty} f\left(\frac{\pi}{a}n\right) \frac{1}{i}(t - \frac{\pi}{a}n) \left[e^{ia(t-\frac{\pi}{a}n)} - e^{-ia(t-\frac{\pi}{a}n)}\right] \tag{3.21}$$

that is:

$$f(t) = \sum_{n=-\infty}^{+\infty} f\left(\frac{\pi}{a}n\right) \frac{1}{a(t - \frac{\pi}{a}n)} \left[\frac{e^{ia(t-\frac{\pi}{a}n)} - e^{-ia(t-\frac{\pi}{a}n)}}{2i}\right] \tag{3.22}$$

whence, by Euler formulae:

$$f(t) = \sum_{n=-\infty}^{+\infty} f\left(\frac{\pi}{a}n\right) \frac{\sin a(t - \frac{\pi}{a}n)}{a(t - \frac{\pi}{a}n)} \tag{3.23}$$

that concludes the proof. □

The third theorem of Shannon concerns the information capacity of signals. Shannon proves that this capacity is directly proportional to the band W of the signal and is inversely proportional to the noise, according to a specific form.

3.12 Third Shannon Theorem

This theorem gives a limit to the capacity of signals.

Theorem 3.3 (Third Shannon Theorem) *Given signal X of limited band W, affected by a noise described by a random variable Z with gaussian distribution (providing a global signal $Y = X + Z$). The limit of the capacity C that this signal can support, as an information channel, is bounded by:*

$$C = W \lg\left(\frac{P+N}{N}\right)$$

where $P = var(X)$ and $N = var(Z)$ are the powers of the signal and of the noise.

References

1. Shannon, C.E.: A mathematical theory of communication. Bell. Syst. Tech. J. **27**, 623–656 (1948)
2. Rosenhouse, J.: Monty Hall Problem. Oxford University Press (2009)
3. Cover, T., Thomas, C.: Information Theory. Wiley, New York (1991)
4. Feller, W.: An Introduction to Probability Theory and Its Applications. Wiley, New York (1968)
5. Carnot, S.: Reflections on the Motive Power of Heat (English translation from French edition of 1824, with introduction by Lord Kelvin). Wiley, New York (1890)
6. Schervish, M.J.: Theory of Statistics. Springer, New York (1995)
7. Brillouin, L.: Scienze and Information Theory. Academic, New York (1956)
8. Manca, V.: Infobiotics: Information in Biotic Systems. Springer, Berlin (2013)
9. Sharp, K., Matschinsky, F.: Translation of Ludwig Boltzmann's paper "On the Relationship between the Second Fundamental Theorem of the Mechanical Theory of Heat and Probability Calculations Regarding the Conditions for Thermal Equilibrium." Entropy **17**, 1971–2009 (2015)
10. Brush, S.G., Hall, N.S. (eds.): The Kinetic Theory of Gases. An Anthology of Classical Papers with Historical Commentary. Imperial College Press, London (2003)
11. Schrödinger, E.: What Is Life? the Physical Aspect of the Living Cell and Mind. Cambridge University Press (1944)
12. Manca, V.: An informational proof of H-Theorem. Open Access Lib. (Modern Physics) **4**, e3396 (2017)
13. Wiener, N.: Cybernetics or Control and Communication in the Animal and the Machine. Hermann, Paris (1948)
14. Bennett, C.H.: Time/space trade-offs for reversible computation. SIAM J. Comput. **18**, 766–776 (1989)
15. Nielsen M.A., Chuang I.L.: Quantum Computation and Quantum Information. Cambridge University Press (2000)
16. Landauer, R.: Irreversibility and heat generation in the computing process. IBM J. Res. Dev. **5**, 183–191 (1961)
17. Bérut, A.: Experimental verification of Landauer's principle linking information and thermodynamics. Nature **483**, 187–189 (2012)
18. Bennett, C.H.: Logical reversibility of computation. IBM J. Res. Dev. **17**, 525–532 (1973)
19. Bennett, C.H.: Notes on the history of reversible computation. IBM J. Res. Dev. **32**(1) (1988)
20. Bennett, C.H.: Notes on Landauer?s principle, reversible computation, and Maxwell's Demon. Stud. Hist. Philos. Mod. Phys. **34**, 501–510 (2003)
21. Manca, V.: A note on the entropy of computation. In: Graciani, C., Riscos-Núñez, A., Pun, G., Rozenberg, G., Salomaa, A. (eds.) Enjoying Natural Computing. Lecture Notes in Computer Science, vol. 11270. Springer, Cham (2018)

Chapter 4
Informational Genomics

Introduction

In this chapter, we consider many aspects of the mathematical and computational analysis of genomes. The presentation is mainly based on books and research papers developed by the authors and collaborators in the last ten years. In particular, we mainly refer to [1–12].

The **genomic alphabet** of characters for **nucleotides** or **monomers** is $\Gamma = \{a, c, g, t\}$, where $\Gamma^\star$ is the set of all possible strings over Γ. A genome $\mathbb{G}$ is a "long" string over Γ of length usually greater than 10^5 (symbols are written in a linear order, from left to right, according to the chemical orientation $5' - 3'$ of Carbon bonds in the pentose of DNA molecules).

The double string structure of genomes has three main reasons:

(1) it guarantees the efficiency of the template-driven duplication in linear time with respect to the length of replicated string;

(2) the transcription (in a single RNA strand) of a genome segment is controlled and requires that a portion of the double strand could be opened, by avoiding free access to the information contained in the genome, which is accessible only in specific situations. At the same time, the double strand provides a structure where better conservation of information is realized;

(3) finally, the double linear structure is the basis of helix DNA conformation, which is essential for its capability to be packed in a very small volume [1, 9].

Virus genomes have lengths from some thousands of monomers to several tens of thousands. Bacterial genomes are long from five hundred thousand (Mycoplasma) to some millions (Escherichia). The genome lengths of multicellular organisms are from hundreds of millions to billions. The human genome is around 3 billion long. Genomes are often organized in parts, called **chromosomes**.

V. Manca and V. Bonnici, *Infogenomics*, Emergence, Complexity and Computation 48,
https://doi.org/10.1007/978-3-031-44501-9_4

For the analyses of genome organization, we can forget double strings, realized by paired DNA strands, because, according to the complementarity principle one strand is entirely implied by the other one.

4.1 DNA Structure

As already noticed, the geometry of the DNA helix allows that very long DNA double sequences could be packed in small volumes (in eukaryotes they are within the nucleus, typically a spherical space of radius around one micron (10^{-6} m). The global length of DNA molecules inside one human cell is around two meters. If we consider that an average estimation of the number of cells in a human body is around 10^{14}, we deduce that the total length of DNA inside a human body is 600 times the distance from the Earth to the Sun. This means that the problem of DNA compression is crucial for the existence of life.

The essence of this structure can be explained in terms of monomeric triangles. In fact, any nucleotide is a monomer A of a polymer, concatenated with the next monomer B of the concatenation line to which it belongs and paired with the corresponding complementary monomer C of the paired strand. Abstractly this provides a triangle ABC. In order to reach a coherent and stable structure the angles of monomeric triangles need to remain almost stable along the whole structure (phosfodiester bonds along the concatenation line and hydrogen bonds along the pairing line). This means that the wrapping of DNA double strands must guarantee the uniformity of monomeric triangles. For this reason, any spiral where concatenated triangles are arranged along the same plane is forbidden, because in that case, the concatenation angles should vary according to the growth of the spiral radius. In DNA helix, concatenation is arranged within an ideal cylinder (Fig. 4.1) of a radius of 1 nanometer and a height of 3.4 nm, corresponding in the most typical DNA form, called B-form, with a 2π rotation along the axis realized by 10, 5 paired nucleotides (Fig. 4.5) and a concatenation rotation angle of about $\pi/5$, around the cylinder axis (Fig. 4.3).

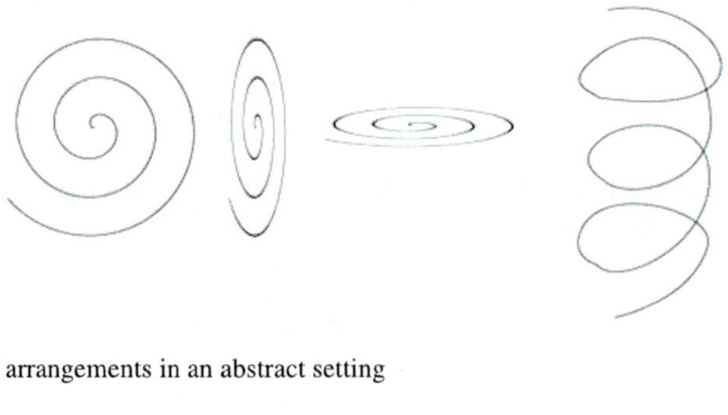

Fig. 4.1 Spiral arrangements in an abstract setting

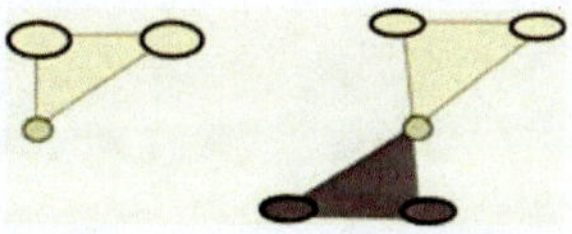

Fig. 4.2 A single monomeric triangle and two paired triangles

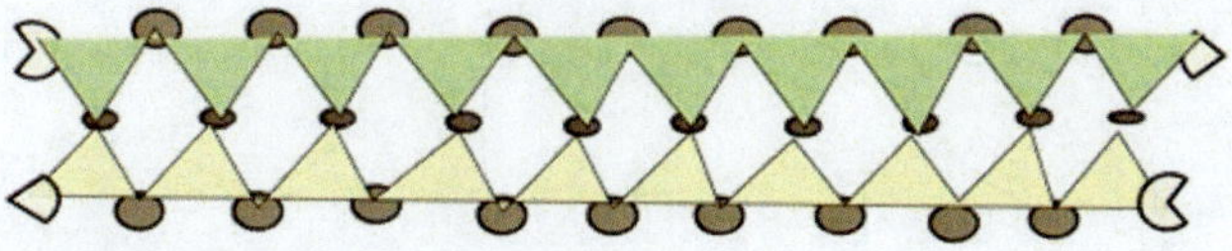

Fig. 4.3 A double line of paired monomeric triangles

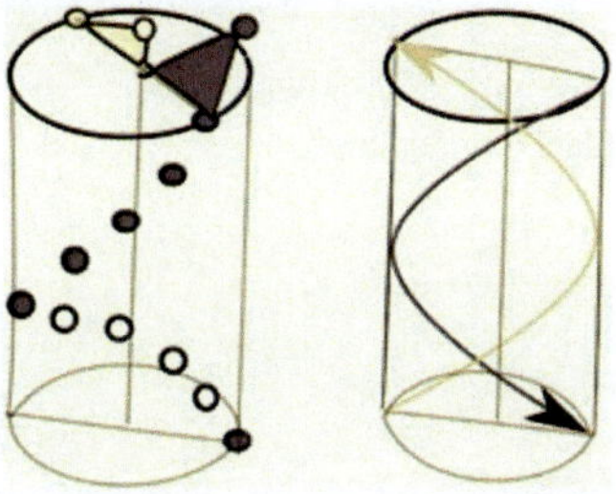

Fig. 4.4 Paired monomeric triangles

Double helix arrangement of monomeric triangles guarantees constant concatenation angles and provides a very stable structure (pairing bonds are internally closed and protected (they are very weak and hydrophobic), being the basis of other wrapping levels that provide an incredible compression of the bilinear structure (Figs. 4.2, 4.3, 4.4 and 4.5).

The helix height has a reduction, with respect to the linear arrangement, which can be evaluated by placing the cylinder's lateral surface on a plane. In this way, the line of concatenated nucleotides results to be placed along the diagonal of a triangle having an edge of 2π nanometers and the other edge of $3, 4$ nm (Fig. 4.6). The length compression ratio ρ of helix arrangement with respect to the concatenation line of a complete rotation around the cylinder axis is thus:

$$\rho = 3.4/\sqrt{3.4^2 + 6.28^2} = 0, 429\ldots$$

Double helix structure gives a reduction of about the 40% of the original length, but, most important, the cohesion of the helix arrangement permits the other four folds of the DNA sequence [13] (nucleosomes, loops, fibres, chromatids).

Monomers are **chiral** objects, that is, 3D structures where no plane can divide them into two symmetric parts (each one a mirror image of the other one). The notion of chirality was introduced by Lord Kelvin (William Thomson the physicist who gave importantly contributes to thermodynamics) is essential for DNA structure and expresses a property of hands (chiros in Greek means hand). An example of an

Fig. 4.5 The double helix of monomeric triangles

Fig. 4.6 The helix cylinder placed on a plane

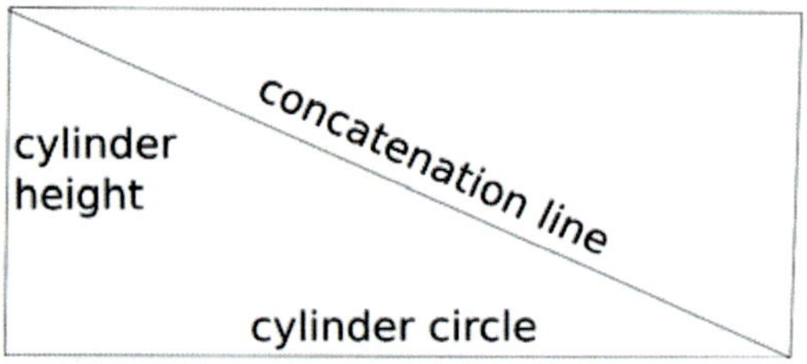

object that is not chiral (achiral) is a bi-glove, which a right hand, or indifferently, a left hand can wear, because is identical to itself after a rotation around the axis passing through the middle finger.

A more general definition of chirality is based on group theory and permits to extend chirality in metric spaces which are not Euclidean. However, the usual definition in 3D Euclidean space is adequate for our present discussion. Two objects formed by parts that are in a reciprocal correspondence (head, body, tail, ...) are homochiral when the cartesian coordinate systems assigned to them by means of the spatial relations among their parts result to have the same handedness: left-hand or right-hand. This means that, if x and y are two axes of the related cartesian coordinate system, an x-to-y rotation (through the convex xy angle) results anti-clockwise, with respect to the z-axis, in a left-handed system, while results clockwise in a right-handed system.

As we will explain monomers are homochiral (with the same chirality), but also paired helices, forming the DNA double helix, must be homochiral, otherwise, they would cross repeatedly on the surface of the cylinder. The most stable form of DNA is the B form, which is a right-handed double helix.

DNA helix follows precise logical and geometric principles making this structure a sort of necessary consequence of them. The basic principles are given by the following items.

A wider system of principles will be given at the end of this section [1].

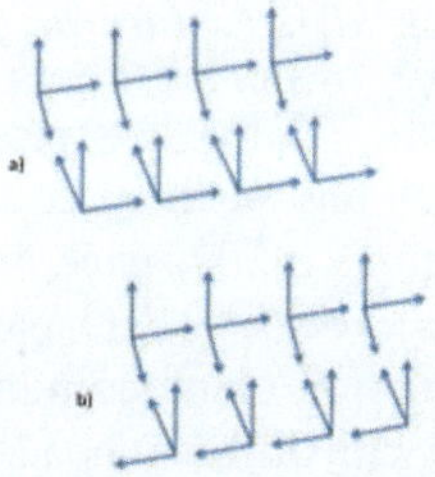

Fig. 4.7 The two possible arrangements of cartesian systems of nucleotides in a double strand. In the arrangement (**a**) nucleotides are concatenated in the same direction. In the arrangement (**b**) nucleotides are arranged in an anti-parallel way. In case of **a** paired monomers have different chiralities, while in case **b** all monomers have the same chirality

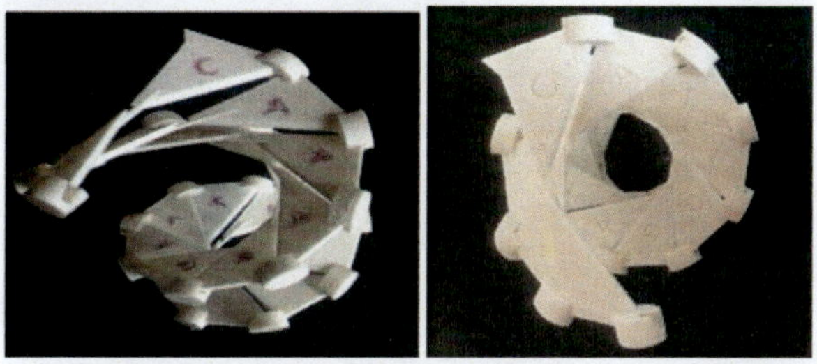

Fig. 4.8 Two vistas of a double helix of monomeric triangles produced with 3D printer [1]. Angles conform with those of B-DNA

- Double linearity: a double linear structure provides a **template driven sequential copy** process in a time that is **linear** with respect to the length of the duplicated sequence.
- Complementarity: In order to obtain two single linear templates from a double strand, the chemical **pairing bonds are weak** (hydrogen bonds). This possibility is better obtained by different (complementary) molecules because bonds between molecules of the same type are covalent bonds that are strong.
- Anti-parallelism: The chirality of a monomer is given by: the x-axis, or concatenation axis, going from the tail to the head; the y-axis, or pairing axis, going from the middle point of the tail-head line to the corresponding point of the paired monomer; z-axis orthogonal to the other axes, with one of two possible orientations, depending on the handedness of chirality. **Monomers are homohiral**, hence, in both strands, they have the same chirality, otherwise, for each nucleotide, we would have the left-hand and the right-hand version, with a complication of the double strand structure and of the related process of duplication. **Pairing is specular**, so that monomers are oppositely oriented with respect to the pairing axis. In such a way, each monomer can pair with only one monomer. Concatenation goes in opposite directions in the paired strands, otherwise, it is easy to realize (see Fig. 4.7) that any monomer in one strand would have a chirality different from that of its paired monomer, against the assumption of homochirality (Fig. 4.8).

The following propositions emphasize the role of the chirality of monomers.

Proposition 4.1 *Chirality of the 3D-structure of monomers is necessary for the correct formation of bilinear DNA strands.*

Proof If monomers were achiral, any of them could be brought to coincide with its mirror image by roto-translations in 3D space, with respect to any plane. Since template-driven duplication is a process where pairing bonds are formed before concatenation bonds, the free motion of monomers getting close to the template could lead to pairing a monomer with different head-tail directions without affecting the orientedness of the pairing direction. However, neither head-to-head nor tail-to-tail concatenation bonds are allowed to form the DNA double strands. □

Proposition 4.2 *Paired monomers are homochiral if and only if their concatenation directions are antiparallel in the paired strands.*

Proof The requirement of specular pairing (see Fig. 4.7) entails that, if paired monomers would concatenate in the same direction, then they would be mirror images of each other, whence they would have different chirality. Conversely, specular pairing and anti-parallelism together imply homochirality, because it is easy to realize, that a roto-translation of the paired monomer makes chiralities of two paired monomers coincident. □

Let us give ten structural principles of DNA structure, putting in evidence the logic motivating them.

1. Directed linearity (of concatenation)
2. Specularity of pairing (no more than two monomers may pair)
3. Complementarity of pairing (weak pairing)
4. Monomeric chirality (unique pairing direction)
5. Homogeneity of chirality (Monomer can belong to both paired strands)
6. Antiparallelism of concatenation (implied by conditions 2 and 4)
7. Helix folding (uniformity concatenation angles)
8. Asymmetry of grooves (phase angle $< \pi$)
9. Bistability of helix (winding-unwinding)
10. Silencing of portions (hiding substrands)

Let us summarize all together the motivations of these principles (by repeating some already given explanations):

Directed linearity is necessary to realize a structure that is a text, composed of letters, with a reading direction.

Specularity of pairing ensures a pairing that "closes" the text making information inaccessible, and more protected.

Complementarity of pairing is related to three main aspects: (a) it is the basis of the template-driven copy of DNA strands: (b) it permits the closure of DNA texts in a reversible way because complementary bonds are weak (with respect to the concatenation bonds), and closed text can be opened (enzymes, called helicases, are devoted to this operation) and closed again; (c) pairing bonds are placed in the middle between the two paired strands, whence inaccessibility and protection are realized at the same time.

Monomeric chirality ensures that pairing happens correctly with respect to the concatenation direction.

Homogeneity of chirality ensures that all monomers have the same chirality, and can be placed indifferently in any strand. If two chiralities for each type of monomer were possible, according to the specularity, paired monomers would have different chirality and the same concatenation direction in both paired strands, by having a greater complexity in the overall elaboration of DNA texts. Therefore, **Antiparallelism of concatenation** is a consequence of homogeneity and chirality.

Helix folding guarantees equal concatenation angles in different positions of a strand that needs to be folded for reducing its spatial volume.

Asymmetry of grooves refers to the agents elaborating DNA helix. In fact, the major grove is the part where agents enter for performing their actions, as is shown in the figures, the asymmetry of groves establishes a preferential way of accessing to the helix.

Bistability of helix is a crucial aspect for the correct use of information. In fact, information needs two states: (a) the inactive state, when it is conserved, and (b) the active state when it is operative and expresses statements executed by specific agents in some cellular contexts. The mechanism of closing and opening portions of paired strands is further controlled by the helix conformation. In fact, Helix formation consists of a winding of the double strand according to a rotational direction, determining a screw tightening, usually clockwise oriented, with respect to the cylinder axis of the helix. When a portion of the helix is unwinded, then strands are separated for access to their information. Therefore, the helix form becomes a further mechanism for controlling the bistability between the states (a) and (b). This aspect is essential because information becomes useful in directing dynamics only if it is used at the right moment, otherwise, it could be useless or even dangerous to the system. In this concern, DNA appears as a mere container of biological information. The activation of information is realized by means of DNA **transcription**, where a portion of one DNA strand is transformed, by means of suitable enzymes, into an RNA strand (U monomer replaces T). Then, biological actions are performed by RNA, which is a single-strand polymer with a short life, sufficient to realize specific actions inside the cell. RNA, probably, was the polymer at the origin of life, and by a process of **translation**, an RNA strand is transformed, according to the **genetic code** (defined on triplets of RNA letters A, U, C, G) into a sequence over an alphabet of 20 letters (amino-acid molecules). These sequences, when formed, assume specific spatial configurations, giving them the possibility of realizing specific biological functions. These biomolecules are called **proteins**, and the process of their realization is called **protein synthesis**. Thus, proteins are **conformational** and **functional** biomolecules, and constitute fundamental biological agents. However, RNA single strands, being single-stranded, are subjected to the attraction between different portions of the same strand, based on complementarity. For this reason, they have the ability to fold, assuming configurations and consequent capacities of realizing specific functions. In this sense, RNA strands are informational and functional molecules, at the same time. This is a natural condition for RNA strands being most probably the first biopolymers, at the beginning of life, and also a natural condition for being the intermediaries between DNA and proteins.

Silencing of portions is a radical way to close some portions of DNA texts. It is the genomic basis of multicellularity. This phenomenon corresponds to the idea of a **super-genome**, that is, a genome that originates many genomes (one for any kind of cell, in an integrated system of different cells). This permits to the maintenance of an original genome, by diversifying it in several forms. A toy example can clarify the main idea. Let $ABCDEFG$ a partition of a string in 7 parts, then let indicate inside brackets silenced parts: $A[B]CD[E]F[G]$, $[A]BC[D]E[F]G$, $AB[CD]E[F]G$, that are equivalent respectively to $ACDF$, $BCEG$, $ABEG$. In this way germinal cells, those responsible for the reproduction of the whole organism, have the global genome without silenced parts but, in the course of the formation and development of a new organism, cells of different kinds (depending on their relative localization in an initial population of equal amorphous cells) receive the genomes with silenced parts, each of them developing functions that are proper of their specific type (around 250 main types in the human body).

4.2 Genome Texts

From the point of view of genome functional organization, we can distinguish seven kinds of regions (see, for example, [14] for specific analyses):

- Transcriptional-Translational regions (genes);
- Pure Transcriptional regions related to different types of RNA modulation activities;
- Regions of promotion/activation where factors and co-factors (proteins) adhere;
- Regulatory regions related to protein coordination functions (homeobox);
- Regions with topological functions related to wrapping/unwrapping activities;
- Regions hosting DNA recombination hotspot;
- Regions related to evolutionary remains, without no direct current functional meaning, repositories in a standby situation.

The classification above does not pretend any completeness, it wants only to point to a very general functional organization based on the main aspects of genome functions. Moreover, only large regions are considered. Of course, small pieces, from some hundreds to some thousands of nucleotides (introns, exons, transcription starts, and so on) surely provide a great number of categories.

Let us consider a very simple numerical argument showing that a complex and articulated mechanism of categorization underlies the assembly of small strings. All the possible strings of length 100 over four letters are 4^{100}, that is, more than 10^{60}. This number is thousands of times greater than the number of atoms on Earth. Therefore, the DNA strands of length 100 that are present in real genomes provide an infinitesimal fraction of all the possible strands of that length. This means that these strands are selected according to very specific constraints. The discovery of some of the main principles underlying this selection would be a great step in understanding genome structures. It is very reasonable to postulate that these principles exist and that mathematical and computational analyses are possible ways for finding them. The next section about genomic dictionaries is based on this assumption, and in that context, some formal categories of genomic words will be considered.

Genomes direct the basic activity of cell metabolism and all the major cell functions are mediated by genes that encode proteins. Proteins are effectors of functions encoded by genes. However, this simple schema works with a complex network of functions, with different levels, that involve, genome segments, generalizing the notion of genes, RNA strands of different types, and proteins. The interactions among them determine a circularity, because proteins that are produced by genes are at the same time, in many phases, crucial for directing genome activities (for example, by activating, promoting, regulating, or inhibiting transcriptions). This circularity is a sort of chicken-egg paradox, which can be solved only at the evolutionary level, postulating protein-like molecules (short amino-acid chains) before the existence of the protein synthesis process, and proto-genomes (most probably based on RNA) before DNA standard genomes [15].

Genomes play a crucial role in transmitting information along life generations. In this sense, they have a double nature: (i) they are the operating systems of cells, (ii) they summarize the evolutionary passages of the biological information producing the species, that is, the history of the lineages from the origin of life up to the formation of organisms hosting them. The analogy with computer operating systems is very attractive, but of course, can be misleading for several reasons. In fact, genomes include pieces that activate and regulate processes, but the way these processes are integrated is not comparable, in many respects, with those acting on electronic computers. First, genomes are self-assembly products, and the processes that they promote are autonomous. For this reason, cooperation is dictated by no central unit but emerges as a result of a network of interactions where the only common attitude is the reproduction of the system hosting the whole dynamics.

We could say that life is a process that wants to keep living. This implies two auto-referential strategies: maintaining life processes in time and, at the same time, propagating them in space. Reproduction is the solution to both these requirements because given the spatial and temporal finiteness of any organism, required by its physical realization, reproduction is a way for overcoming individual limitations in space and in time. Therefore, genomes are intrinsically replicable and probably, in the history of life, organism reproduction is a consequence of searching for correct genome replication. In any case, genomes are individual structures that can be fully analyzed in the context of their evolutionary nature, where individuals are only instances of genealogical lines from which species arose.

Genomes have a double nature, directing the life of individuals, and propagating along evolutionary lines through species. For this reason, they memorize the species' genealogy but determine also the changes aimed at exploring new life possibilities. In fact, just at the moment of their transmission, they include variability mechanisms.

These are intrinsic to their nature, in the sense that even if they search for the best fitting of individuals with the environment, they in any case change even without any external pressure, but just for a free intrinsic mechanism of casual exploration. They are subjected to computational rules for their functioning inside cells, but they are subjected to random events on which natural selection acts for fixing the most advantageous for better adaptability to the environment where organisms are placed.

For this reason, a mathematical and computational analysis of genomes has to cope with their computational ability (see [16, 17] for analyses in a DNA computing perspective), and at the same time with random processes. A clear and rigorous account of the way genomes balance and dominate both aspects is surely one of the most difficult challenge to a deep understanding of the nature and dynamics of genomes.

Three were the factors of Darwin's theory of evolution: **heredity**, **variability**, and **selection**. The discovery of DNA and genomes not only do not contradict Darwin's perspective but give a new comprehension of evolutionary dynamics at a molecular level.

Heredity is the mechanism of passing biological information by duplicating the genome of the parent cell and leaving that the copy becomes the genome of the generated cell. However, in the passage, some variability is included due to casual mutations, and copy errors, but even specific mechanisms of variability in sexual reproduction reach the level of a very specific recombination machine: from the two genomes of parents a new recombined genome is obtained with parts coming from the two parental cells. The key point of Darwin's formulation is that information does not pass from individuals to the species, because only their genomes matter, nothing else about the individual history of the parents.

As already noticed, genomes vary for a free tendency of exploring new life possibilities. This capability of genomes is the main mechanism for realizing the fitness of the organisms to the environment where they live. Genome changes, in their extreme details, are representable by mechanisms of string rewriting. Each genotype is associated with a phenotype. Natural selection favourites phenotypes that increase fitness to the environment. This selection improves the average quality of life of individuals, because the better they live, the greater their probability of reproduction, and their descendants maintain alive the genomes that they received from their parents. In this way, the genomes of organisms more adapted become dominant and the corresponding genomes have a greater probability of passing to the species of their evolutionary path. In this way, information passes with no direct intervention in the individual life. Therefore, individual genomes are containers of biological information. a substance flowing through them, and mixing their contents, by recombination mechanisms. The only intervention of the individual in this process is dying or surviving. Death is a necessity of evolution.

A classical example of evolution is that of a giraffe. This animal obtained his long neck only by chance, and surely not for the attempts to eat leaves of high trees. On the contrary, the advantage of a casual long neck became an advantage in an environment with high trees of good leaves on the top. In general, species evolve, toward increasing fitness, because some individuals change, casually, in a direction that is advantageous with respect to their environment, Therefore, organisms evolve because they change, while they do not change for evolving.

In this evolutionary scenario, randomness is the king of the process, and paradoxically it is a necessity. Deep mathematical properties of randomness ensure great efficiency in exploring spaces of possibilities. In particular, a property, called **ergodicity**, can be intuitively visualized by a dynamics visiting, homogeneously in time, all the regions of a space of possibilities. In this way, if optimal solutions to a given problem exist, then they are surely found.

However, the mechanism of fixing some advantageous steps, and integrating the single steps toward a coherent structure which, in the end, appears as a sort of performed project remains a mystery, in terms of our actual knowledge about genomes.

About species formation, how can we measure the steps necessary for passing to a new species from a previous one? In fact, changes cumulate along generations, but at some point, speciations become a jump, where the genomic differences result in a radical change of the play, and the resulting genome "speaks" a different language with respect to those of genomes of its generative lineage. This process is very similar to the passage from one linguistic system to another one. Also, in this case, a new language is generated that results incompatible, for communication, with the language from which it stems. When and why this incompatibility arises? In the sexual species, incompatibility means the impossibility of procreation from sexual mating among individuals of different species. But again, when and why does an individual represent the birth of a different species with respect to that his/her parents?

The actual impossibility of rigorous answers to the previous question is almost centred on the absence of quantitative laws describing the evolutionary dynamics. These laws are the core of any deep future comprehension of the basic mechanisms of life.

4.3 Genome Dictionaries

Given a genome $\mathbb{G}$ its complete dictionary $D(\mathbb{G})$ is given by $sub(\mathbb{G})$, that is, by all the strings occurring in $\mathbb{G}$ between two positions:

$$D(\mathbb{G}) = \{\mathbb{G}[i, j] \mid 1 \leq i \leq |\mathbb{G}|\}$$

and moreover:

$$D_k(G) = D(\mathbb{G}) \cap \Gamma^k.$$

where Γ^k is the set of all possible words of length k over the alphabet Γ.

Any subset of $D(\mathbb{G})$ is called a **dictionary** of $\mathbb{G}$. Elements of $D(\mathbb{G})$ are also called **words**, **factors**, **k-mers**, **k-grams** of $\mathbb{G}$. A prefix k. generic or of a given value, is used when we want explicitly mention the length of strings. We will write $\alpha \in \mathbb{G}$ as an abbreviation of $\alpha \in D_{\mathbb{G}}$.

The set of positions where (the first symbol of) a string α occurs in $\mathbb{G}$ is denoted by $pos_{\mathbb{G}}(\alpha)$, and the **multiplicity** of α in $\mathbb{G}$, that is, the number of times it occurs in $\mathbb{G}$, is indicated by $mult_{\mathbb{G}}(\alpha)$.

Let us define some classes of strings that are very useful in the analysis of genomes:

1. **Repeats** are words with multiplicity greater than 1.
2. **Hapaxes** are elements with multiplicity exactly equal to 1.
3. **Duplexes** are elements with multiplicity exactly equal to 2.
4. **Nullomers** are the shortest strings that do not occur in the genome.

The term **hapax** comes from philology and means a word occurring in G once (from a Greek root for once). A **repeat** of $\mathbb{G}$ is a word occurring in $\mathbb{G}$ at least twice. A **maximal repeat** is a repeat that is not a substring of another repeat. A **minimal hapax** is an hapax where removing its first symbol, or its last symbol, produces a repeat. Equivalently, elongating a maximal repeat α with a letter $x \in \Gamma$, in strings αx or $x\alpha$, provides minimal hapaxes.

The maximality of repeats can be considered in two different senses. Let α, β be repeats of $\mathbb{G}$ while α is a prefix of β and β is a maximal repeat, then α is not a maximal repeat. But, if α occurs in $\mathbb{G}$ in some positions where β does not occur, and in these positions, α is not a substring of any repeat of $\mathbb{G}$, then α is a maximal repeat, with respect to these positions. Then, it is useful to consider the notion of **positional maximal repeat** as a repeat that is maximal with respect to some non-empty set P of positions, called the set of **maximal occurrences** of α in $\mathbb{G}$.

If we consider texts in natural languages, words such as *mean, meaning, meaningful, meaningfulness* result to be, in some English texts, cases of positional maximal repeats.

The notion of positional maximal repeat is very close to the usual notion of *word* in natural languages. Conversely repeats that are never positional maximal repeats, can occur only internally to positional maximal repeats corresponding to *roots* (in a similar sense as *mean, ing, ful, ness* are English roots). The notion of repeat provides a hierarchy of units: *letters, roots, words, syntagmas, statements, composite statements*.

Of course, an hapax of $\mathbb{G}$, does never correspond to a word, because it occurs once in $\mathbb{G}$. But, if we consider a class of genomes, or in general a class of texts, then the hapaxes of these texts can be repeats in the union of all the texts of the class.

If the length of $\mathbb{G}$ is n, that is, $|\mathbb{G}| = n$, then for $k > \log_k n$ surely exist k-mers that are nullomers of $\mathbb{G}$, and increasing k, the most part of k-mers surely are nullomers, because in this case $4^k > n$, therefore surely there are k-mers that cannot occur in $\mathbb{G}$. This means that for k around the logarithmic length of $\mathbb{G}$ is realized the selection of k-mers occurring in $\mathbb{G}$, and in this selection is expressed the biological meaning of k-mers and of the more complex syntactical units where they occur.

All the set-theoretic operations and relations of formal language theory can be applied to dictionaries. In particular, given a dictionary L, then $|L|$ is the number of its elements (with notation overloading because $|\alpha|$ also denotes the length of the string α).

Interesting dictionaries of $\mathbb{G}$ are those that covers $\mathbb{G}$ but consisting of a relatively small numbers of words with respect to the length of $\mathbb{G}$.

We say **absolute dictionary** of $\mathbb{G}$, denoted by $abs(\mathbb{G})$, the dictionary constituted by the positional maximal repeats and minimal hapaxes of $\mathbb{G}$. This notion of a genome dictionary is not related to a fixed length of words, and in many aspects seems to reflect intrinsic characters of a genome.

Proposition 4.3 *For any k, $|D_k(\mathbb{G})| \leq |\mathbb{G}| - k + 1$ and there exist genomes for which equality holds.*

Proof In fact if we move forward a window of length k from the first position of $\mathbb{G}$, then the rightmost position where the window can be placed is the position $|\mathbb{G}| - k + 1$. The existence of genomes having the maximum number of possible different k-mers follows from the existence of de Bruijn genomes, which will be presented in the next section. □

Proposition 4.4 $|D(\mathbb{G})| < |\mathbb{G}|(|\mathbb{G}| + 1)/2$.

Proof There is one word of length $|\mathbb{G}|$, at most two different words of length $|\mathbb{G}| - 1$, at most three different words of length $|\mathbb{G}| - 2$, and so on. But words of length $1 = |\mathbb{G}| - (|\mathbb{G}| - 1)$ are less than $|\mathbb{G}|$ (unless $|\mathbb{G}| = 4$). Therefore according to Gauss' summation ($\sum_{i=1}^{n} = \frac{n(n+1)}{2}$), we get the given upper bound. □

4.4 Genome Informational Indexes

Informational indexes of a genome are parameters related to the global properties of genomes and are defined by means of particular genomic distributions and dictionaries.

The k-mer distribution of $\mathbb{G}$ is given by the frequency in $\mathbb{G}$ of each k-mer of dictionary $D_k(\mathbb{G})$.

The entropy defined by the genomic k-mer distribution is called the k-entropy of $\mathbb{G}$ and is denoted by $E_k(\mathbb{G})$.

The following indexes are of particular importance.

1. $LG(\mathbb{G}) = \lg_4(|\mathbb{G}|)$, the **logarithmic length** of $\mathbb{G}$.
2. $mrl(\mathbb{G})$ the **maximum repeat length**, that is, the maximum length of maximal repeats.
3. $mcl(\mathbb{G})$ the **maximum complete length** such that all possible strings of that length occur in $\mathbb{G}$.
4. $mfl(\mathbb{G})$ the **minimal forbidden length** such that surely strings of that length exist that do not occur in $\mathbb{G}$.
5. $mhl(\mathbb{G})$ the **minimum hapax length**, that is, the minimum length of minimal hapaxes.
6. $mgl(\mathbb{G})$ the **minimal hapax greatest length**, that is, the value of the maximum among the lengths of minimal hapaxes.
7. $\nabla(\mathbb{G})$ the **no-repeat length**, that is, $mrl(\mathbb{G}) + 1$.
8. $\Delta(\mathbb{G})$ the **no-hapax length**, that is, $mhl(\mathbb{G}) - 1$.

9. $LX_k(\mathbb{G})$, the **average multiplicity** of k-mers, called k-**lexical index**
10. $E_k(\mathbb{G})$, the k-**entropy**
11. $LE(\mathbb{G})$, the **logarithmic entropy**, that is, k-entropy for $k = \log_4 n$, where n is the length of $\mathbb{G}$.
12. $2LE(\mathbb{G})$, the **binary logarithmic entropy**, that is, k-entropy for $k = \log_2 n$ (the interest of this value will be clarified later on).

Other indexes that could be interesting in some contexts are **Repeat indexes** giving the lengths and the numbers of maximal repeats or the lengths and the numbers of maximal positional repeats (minima, maxima and average values).

Any genomic dictionary determines indexes given by the maximum, minimum, and average multiplicities of its k-mers.

Some simple facts follow from the definitions given above (when the genome is understood indexes will be used without explicit indication of the genome. For example LG abbreviates $LG(\mathbb{G})$).

Proposition 4.5 *In any genome, a string including an hapax string is an hapax too.*

Proposition 4.6 *In any genome, a string included in a repeat string is a repeat too.*

The **right elongation** of a string α is any string of which α is a prefix, analogously a **left elongation** of α is a string of which α is a suffix. A string that is the right or left elongation of α is an **elongation** of α.

Proposition 4.7 *Nullomers are elongations of minimal hapaxes.*

Proof In fact, if an hapax α occurs in a genome, then all the strings of the genome including α are hapaxes too. Let $\beta\alpha\gamma$ be any string including α in the given genome. Then, all the strings where β and γ are replaced, respectively, by strings different from them are surely strings that cannot occur in the genome. □

Proposition 4.8 *In any genome, a string longer than mrl is an hapax, and any string shorter than the mhl is a repeat.*

The interval $[\Delta(\mathbb{G}), \nabla(\mathbb{G}])$ is called **hapax-repeat interval** of a given genome. All the strings longer than $\nabla(\mathbb{G})$ are hapaxes, those shorter than $\Delta(\mathbb{G})$ are repeats.

Proposition 4.9 *In any genome $mcl < LG$.*

Proof In fact, by the definition of mcl it follows that:

$$4^{mcl} \leq n - k + 1$$

therefore:

$$mcl \leq \lg_4(n - k + 1) < \lg_4(n).$$

Proposition 4.10 *In any genome $mcl = mfl - 1$.*

Proposition 4.11 *In any genome $mcl \leq mhl$.*

Proof In fact, If $k = mcl > mhl$, then an hapax α of length $k - 1$ should exist. This means that three among the string αa, αc, αg, αt cannot occur in the genome. But, this means that $k \neq mcl$, which is a contradiction. □

Proposition 4.12 *In all genomes* $mfl \leq mhl + 1$.

Proof Let $k = mfl$, then a k-mer α has to be an hapax, this means that the letter following it is only of one kind. This implies that, for some letter x the $(k + 1)$-mer αx is a nullomer. Therefore, $k = mfl \leq mhl + 1$.

The following proposition gives a chain of inequalities among mcl, mfl, mhl indexes.

Proposition 4.13

$$mcl = mfl - 1 \leq mhl$$

Proposition 4.14 *Maximal repeats have at most multiplicity 5.*

Proof In fact, if α is a maximal repeat no its elongation αx with $x \in \Gamma$ can occur twice, otherwise α would not be maximal. This means that symbols after the occurrences of α must be different. But, if α occurs more than four times, in the occurrences where a symbol occurs after α, this condition cannot be verified. Therefore α has to occur at most four times in the middle of the genome and only once at the end of the whole genome. □

A genome $\mathbb{G}$ is a k-hapax genome when its dictionary $D_k(\mathbb{G})$ consists only of hapax, moreover $\mathbb{G}$ is also a complete hapax genome if $D_k(\mathbb{G})$ consists of all 4^k k-mers. In this case the length of $\mathbb{G}$ is $4^k + k - 1$. The following genome is a 2-hapax genome:

$$aaccggttagatctg$$

The following is a 2-hapax complete genome. It is complete because all pairs occur once, and its length is $17 = 2^4 + 2 - 1$, capable of containing all the possible pairs:

$$aaccggttagatctgca$$

The following genome, where the last two symbols of the previous genome were permuted, is no more an hapax genome:

$$aaccggttagatctgac.$$

A general method for generating hapax genomes will be given in a next section where de Bruijn graphs are introduced. According to the equipartition property of entropy, we have the following proposition.

Proposition 4.15 *Any k-hapax genome $\mathbb{G}$ of length n has the maximum value of $E_k(\mathbb{G})$ among all genomes of the same length.*

In real genomes usually $mhl \leq mrl$, but this inequality does not hold in general. In fact, in the following genome:

$$aattaacc$$

the shortest hapaxes are long 3, but the longest repeats are long 2, therefore $mhl > mrl$.

If a k-mer is an hapax, it is univocally elongated, but the converse does not hold, because a k-mer can elongate in only one way, without being hapax.

4.5 Genome Information Sources

The most important notion of informational genomics is that of **genomic distribution**. It is defined by a variable $X_{\mathbb{G}}$ taking values in a set of components/features of $\mathbb{G}$ (positions, k-mers, substrings, distances, substring occurrences, ...). Counting the number of times that an event $X_{\mathbb{G}} = x$ occurs provides a multiplicity, and consequently, a frequency associated to the value x, whence a probability distribution for $X_{\mathbb{G}}$. A genomic distribution is given by a pair (X, p), where X is a variable, defined over a genome or over a class of genomes, and p is a function assigning a probability $p(x)$ to each value x assumed by X. The pair (X, p) is also an information source in the sense of Shannon.

Given a genomic sequence $\mathbb{G}$, several types of functions and distributions can be defined on it, depending on the different perspectives from which genomic sequences are investigated.

The **position** function $pos_{\mathbb{G}}$ assigns to each word $\alpha \in D(\mathbb{G})$ the set of positions where α occurs in $\mathbb{G}$. Formally:

$$pos_{\mathbb{G}}(\alpha) = \{i : \mathbb{G}[i, i + |\alpha|] = \alpha\}.$$

The **word multiplicity** function $mult_{\mathbb{G}}$ maps each word α into its multiplicity $mult_{\mathbb{G}}(\alpha)$, within the given genome:

$$mult_{\mathbb{G}}(\alpha) = |pos_{\mathbb{G}}(\alpha)|.$$

The **word length** distribution assigns to each positive integer $k \in [1 \ldots |\mathbb{G}|]$ the number of distinct words of $|D_k(\mathbb{G})|$ that occur in the genome $\mathbb{G}$.

The value $lcl(\mathbb{G})$, called the **limit cardinality length**, is the shortest length for which words of $\mathbb{G}$ of length $lcl(\mathbb{G})$ are, in number, more than words of $\mathbb{G}$ shorter than $lcl(\mathbb{G})$, but at the same time are, in number, not less than words of $\mathbb{G}$ longer than $lcl(\mathbb{G})$.

The **hapax length** distribution assigns to each word length k the number of hapaxes of such a length:

$$|\{\alpha \in D_k(\mathbb{G}) \mid mult_{\mathbb{G}}(\alpha) = 1\}|.$$

Analogously, the **repeat length** distribution assigns to each word length k the number of repeats of $\mathbb{G}$ having that specific length:

$$|\{\alpha \in D_k(\mathbb{G}) \mid mult_{\mathbb{G}}(\alpha) > 1\}|.$$

Distributions can be converted as percentages by dividing the number of words of a given length k by the total number of k-mers occurring in the given genome, that is:

$$freq_{\mathbb{G}}(\alpha) = mult_{\mathbb{G}}(\alpha)/(|\mathbb{G}| - |\alpha| + 1)$$

because $|\mathbb{G}| - |\alpha| + 1$ is the number of all the occurrences in $\mathbb{G}$ of the words with the same length of α.

Given a distribution over a set A of elements, the associated **Zipf's curve** is obtained by ordering (in increasing or decreasing order) the elements of A with respect to their values of frequency, where elements with the same frequency are also said to be of the same **rank**, and then by associating to the elements of a given rank the corresponding value of frequency.

The **co-multiplicity** distribution assigns to each value m of multiplicity the set of words (or the number of words) of $\mathbb{G}$ having that multiplicity:

$$\{\alpha \in D(\mathbb{G}) \mid mult_{\mathbb{G}}(\alpha) = m\}|.$$

Multiplicity and co-multiplicity distributions are shown to be distinctive features for genomic sequences belonging to different species and phyla [18].

The **entropy length** distribution assigns to each word length k the empirical entropy calculated on the frequencies of the words in $D_k(\mathbb{G})$. This distribution shows different trends that depend on the specific genomic sequence. This characterization is more evident for values of k in the hapax-repeat interval.

Figure 4.9 shows the distributional trends for a random sequence of 10.000 monomers.

Other interesting distributions, are based on the notions of coverage. There are two main notions of coverage relating genomes and dictionaries: (i) sequence coverage and (ii) position coverage. Informally, they denote, respectively, (i) the set of positions of a genome where are located the words of a given dictionary, and (ii) the set of words of a dictionary such that their occurrences, in a genome, pass trough a given position of it.

Given a dictionary D, the **sequence coverage** of D, for a genome $\mathbb{G}$, is defined as the set of positions of $\mathbb{G}$ for which at least one word in D is placed in them. Formally:

$$poscov_{\mathbb{G},p} = |\{p \mid \exists i, j (i \leq p \leq j) \wedge (\mathbb{G}[i, j] \in D)\}|.$$

Starting from the notion of sequence coverage, one can define genomic distributions assigning to a set of words of $\mathbb{G}$ all the positions of $\mathbb{G}$ covered by the considered set. Thus, given the sets $H_k(\mathbb{G})$ and $R_k(\mathbb{G})$, of hapaxes and repeats of length k in $\mathbb{G}$,

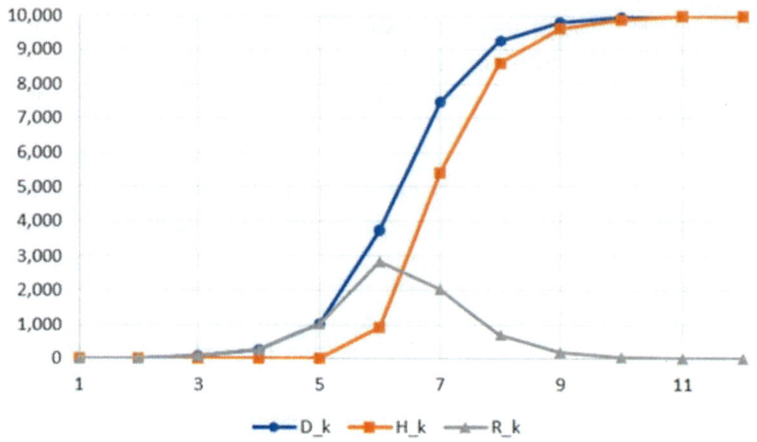

Fig. 4.9 Distributional values of $|D_k|$, $|H_k|$ and $|R_k|$ for a random sequence of 10,000 positions generated on an alphabet of 4 symbols having equal probability. The mhl value equals 6, the maximal repeat length (mrl) is 11 and the minimal forbidden length (mfl) is 6. Moreover, $lg_4(10, 000) = 6.64$

the **hapax coverage distribution** and **repeat coverage distribution** are defined as $k \mapsto |cov_{\mathbb{G}}(H_k(\mathbb{G}))|$ and $k \mapsto |cov_{\mathbb{G}}(R_k(\mathbb{G}))|$, respectively.

In modern sequencing techniques, the notion of **sequencing depth** refers to the average number of *reads* that cover a given position of the sequenced genome. The set of *reads* produced by the sequencing operation can be seen as a set of words, and the sequencing depth corresponds to the notion of **position coverage**. Namely, given a genome $\mathbb{G}$, a position p of it, and dictionary D, the position coverage of p in D is defined as the number of words of D which cover p. Formally:

$$poscov_{\mathbb{G},p} = |\{\alpha \in D \mid \exists i, j(i \leq p \leq j) \wedge (\mathbb{G}[i, j] = \alpha)\}|.$$

The **position coverage distribution** assigns to a value c the number of positions on $\mathbb{G}$ having such a specific value of position coverage. Formally:

$$c \mapsto |\{p : poscov_{\mathbb{G},p}(L) = c\}|.$$

The **Recurrence Distance Distribution** $RDD_{\mathbb{G}}(\alpha)$ is the function assigning to each value d of a distance the number of times two consecutive occurrences of α in $\mathbb{G}$ are separated by a distance d.

It is important to remark that $RDD_{\mathbb{G}}(\alpha)$ is often related to the function of α within the given genome. Namely, an α such that $RDD_{\mathbb{G}}(\alpha)$ assumes a very characteristic pattern, with respect to the other strings, seems to be reasonably considered as an element representing a "genomic word" (see [7] for a detailed analysis of several aspects related to distance recurrence).

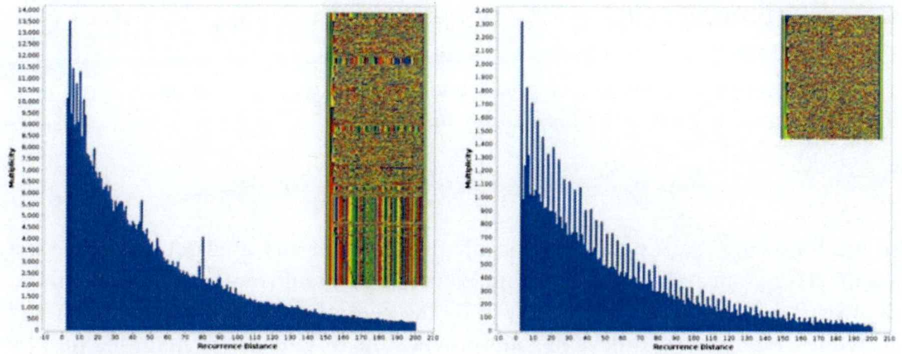

Fig. 4.10 On the left side, recurrence distance distribution (up to distance 200) of the words ATG from the chromosome 22 of Homo sapiens. Besides the distribution curve, there are 4, 081 colored rows, in lexicographic order, of the 84-strings enclosed between the (minimal) recurrences of ATG at distance 81. On the right side, the recurrence distance distribution of the word ATG in *Escherichia coli*. Besides the curve, there are 416 strings $ATG - 78N - ATG$ denoted by coloured rows in lexicographic order, that is, the strings enclosed between the recurrences of ATG at distance 81. The enclosed strings of *Escherichia coli* show low similarity. Contrarily, a great part of the enclosed strings in *Homo sapiens* are similar. Moreover, approximately, the percentage of strings showing similarity corresponds to the portion of the peak emerging from the basic curve

For example, by using RDD it is possible to distinguish regions of genomes where this distribution presents regular peaks that result in strongly correlated with the coding (*versus* non-coding) property of these regions, as it is shown in Fig. 4.10.

In general, the main criterium for discriminating between strings that are putative "genomic words" with respect to strings that do not express such conditions is based on the degree of anti-randomness of RDD. In any long string a substring that occurs randomly has an RDD distribution that approximates an exponential law (is the probability distribution associated with a Poisson process, and usually called waiting time, that is, the number of steps of the process between two occurrences of a rare event). Therefore, any kind of "distance" between $RDD(\alpha)$ and a suitable exponential distribution exp provides a reasonable criterium for characterizing genomic words. The recurrence pattern of a k-mer in a genome is an important aspect related to specific genome features. Entropic concepts applied to RDD distributions can be applied for extracting informational regularities and peculiarities of given genomes or classes of genomes [3–8].

For any genomic distribution, we can immediately define the usual statistical indexes: mean, standard deviation, max, min, and so on.

Many different genomic distributions can be defined for a given genome. However, it is crucial the choice the right variables by means of which genomic information sources are identified that are significant for specific analyses of genomes.

4.6 Genome Spectra

The **k-Spectrum** of a genome $\mathbb{G}$ [19], denoted by $spec_k(\mathbb{G})$, is defined in terms of multiset, that is, a set of pairs *(monomer, multiplicity)*:

$$spec_k(\mathbb{G}) = \{(\alpha, mult_{\mathbb{G}}(\alpha)) \mid \alpha \in D_k(\mathbb{G})\}.$$

In the following, the usual operation of multiset-sum and multiset-difference are assumed for k-spectra. In particular, the elements of a multiset-sum $A + B$ are those of A or B with the multiplicities given by the sum of their multiplicities in A and B, while those of $A - B$ have as multiplicities the differences of multiplicities, by setting zero the negative values. Moreover, any multiset does not change if a new element with zero multiplicity is added o removed, therefore a multiset having only zero multiplicities can be considered equivalent to the empty set.

Assuming all the k-mers in the lexicographic order, any k-spectrum is completely represented by a multiplicity vector W, where the i-component gives the multiplicity, in the k-spectrum, of the k-mer of lexicographic position i.

Now we introduce a number of concepts related to k-spectra that provide very powerful representations of genomes [19].

A genome $\mathbb{G}$ is said to be **k-univocal** (in spectrum) if no genome different from $\mathbb{G}$ has the same k-spectrum of $\mathbb{G}$. The sum of the multiplicities associated to $spec_k(\mathbb{G})$ coincides with $|\mathbb{G}| - k + 1$.

The following toy example of spectra helps to explain the concept. Let us consider the string $a^3b^2a^5$. Its 2-spectrum is the following:

$$spec_2(a^3b^2a^5) = \{(aa, 6), (ab, 1), (bb, 1), (ba, 1)\}$$

moreover:

$$spec_2(a^5b^2a^3) = spec_5(a^3b^2a^5)$$

Therefore, $a^3b^2a^5$ is and $a^5b^2a^3$ are not 2-univocal. However, it is easy to realize that for $k = 5$ we have:

$$spec_5(a^3b^2a^5) \neq spec_5(a^5b^2a^3).$$

Proposition 4.16 *A k-hapax genome is sure k-univocal in the spectrum, but the converse implication does not hold.*

Proof If a genome is k-hapax, then its k-mers can be concatenated in only one way, because they appear only once in the genome. However, the converse implication does not hold. We can show this fact with an example, which for a better reading we give in the English alphabet:

$$canegattogallogattone.$$

All the 4-mers occurring in it are hapaxes, apart from *gatt* and *atto*, which have multiplicity 2. Therefore, the sequence is not 4-hapax. Nevertheless, this sequence is 4-univocal, as can be easily verified. □

Of course, given a genome $\mathbb{G}$, some $k > 1$ has to exist such that $\mathbb{G}$ is k-univocal (surely $\mathbb{G}'$ is univocal for $k = |\mathbb{G}|$). We call **spectrality** of $\mathbb{G}$, denoted by $spl(\mathbb{G})$, the minimum k for which $\mathbb{G}$ is k-univocal:

$$spl(\mathbb{G}) = min\{k \mid \mathbb{G}\ is\ k-univocal\}.$$

If $\mathbb{G}$ is not k-univocal there is at least one different genome $\mathbb{G}'$ having the same spectrum of $\mathbb{G}$. This means that all the k-mers occurrences of the spectrum occur in the two genomes $\mathbb{G}$ and $\mathbb{G}'$ with different orders. A natural question is: *which are the longest substrings occurring in both genomes?*

In the extreme case, these arrangements coincide with the k-mers, but in general, these portions can be longer than k and it would be interesting to find such common components of genomes having the same k-spectrum. Let us introduce some definitions for answering this question in a general setting.

We say that a k-spectrum H **admits** $\mathbb{G}$ when

$$spec_k(\mathbb{G}) \subseteq H$$

where the inclusion has to be intended as multiset inclusion, that is, all k-mers occurring in the left part occur in the right one with an equal or greater multiplicity. Of course, if $\mathbb{G}'$ is a proper segment of $\mathbb{G}$ (prefix, infix or suffix of $\mathbb{G}$), then:

$$spec_k(\mathbb{G}') \subset spec_k(\mathbb{G}).$$

A $(k-1)$-concatenation of two strings $\alpha\gamma$, $\gamma\beta$, also called **concatenation with** $(k-1)$ **overlap**, is the string $\alpha\gamma\beta$ where $|\gamma| = k - 1$. A string α is **univocally** k-**assembled** in a given k-spectrum H, when the k-prefix β of α is in H, and for any proper prefix γ of α, there is only one k-mer of H that can be $(k-1)$-concatenated to it, by right elongating γ of one symbol.

We say k-**spectral segment** of $\mathbb{G}$ any string that is maximal among the univocally k-assembled strings in $spec_k(\mathbb{G})$. A k-**spectral reverse segment** of $\mathbb{G}$ is a k-spectral segment of the reverse of $\mathbb{G}$, and an k-**spectral symmetric segment** of $\mathbb{G}$ is any string that is a k-spectral segment and a k-spectral reverse segment of $\mathbb{G}$.

From the previous definition of k-spectral segment, the following proposition follows.

Proposition 4.17 *A k-spectral segment of $\mathbb{G}$ is also a k-spectral segment of any genome having the same spectrum of $\mathbb{G}$.*

The following proposition easily follows from the previous one.

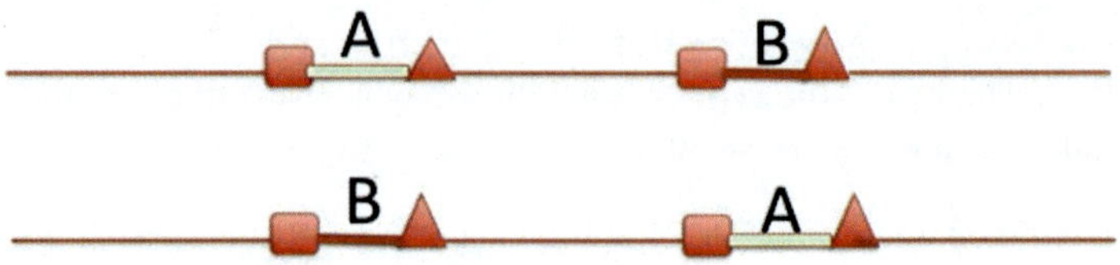

Fig. 4.11 Prefixes and suffixes having the same form denote equal strings long $k - 1$

Proposition 4.18 *Two different genomes* $\mathbb{G}$ *and* $\mathbb{G}'$ *have the same k-spectrum if, and only if, they have the same k-spectral segments, possibly occurring in them with different relative orders.*

We remark that when the condition of univocal elongation does not hold, there are strings that are admitted by the k-spectrum, but that do not occur in the genome.

For example, *canegatto* and *attogallogattone* are the 4-segments of the 4-**spectral segmentation** of

$$canegattogallogattone$$

but *canegattone* is not a 4-spectral segment of that string, even if it is admitted by its 4-spectrum, because *canegatto* is a proper substring of both strings *canegattog* and *canegatton* that occur in the string.

In conclusion, the k-mers of the spectrum of $\mathbb{G}$ occur, in different orders, in all genomes with the same k-spectrum of $\mathbb{G}$, the k-spectral segments of $\mathbb{G}$ extend this property of k-mers of $spec_k(\mathbb{G})$, in the sense that k-spectral segments are the maximal strings, occurring with different orders, in all genomes having spectrum $spec_k(\mathbb{G})$.

Two strings, occurring in a genome $\mathbb{G}$, are spectrally k-permutable when, after they are swapped in $\mathbb{G}$ the k-spectrum remain equal to $spec_k(\mathbb{G})$. Spectral segments determine genomes apart from the swapping of k-spectral segments. It is easy to realize that two strings are spectral k-permutable if they have the same $k - 1$ prefix and the same $k - 1$ suffix. Figure 4.11 shows spectrally k-permutable strings.

If a genome $\mathbb{G}$ is k-univocal, then its k-spectral segments can be concatenated, with $(k - 1)$-overlap, in only one way. However, a genome $\mathbb{G}$ can be k-univocal but overlap concatenations of k-mers of its spectrum may be obtained that do not occur in the genome.

A k-**spectral segmentation** of $spec_k(\mathbb{G})$ (or of $\mathbb{G}$), is the multiset of k-spectral segments of $\mathbb{G}$, each with its multiplicity in $\mathbb{G}$.

The following procedures provide two ways of computing spectral segmentations in two cases: from a given spectrum, and from a given genome.

Algorithm 4.1 (computing a k-segmentation from a k spectrum)
Let H be a k-spectrum. The procedure starts by assigning to the current string w one k-mer chosen out of H and, at each step, searches in H for the k-mer that uniquely elongates w, on the right, or on the left, by updating H after removing the k-mer

occurrence used in the elongation. If more than one possible k-mers of H can be concatenated to both the extremities of the current string, then the concatenation process halts, and the string obtained so far is produced in output, as a k-spectral segment. Then, the procedure restarts with a similar process with another k-mer of H. The procedure halts when H becomes empty. □

In the procedure above it is not essential to establish the choice criterion for the initial string w (the first k-mer in the lexicographic order could be one possibility). Some of the k-spectral maximal segments produced in output during the whole process above can be generated with a multiplicity greater than 1. The multiset of these segments with their respective multiplicities is the k-segmentation of H. Clearly, in the genomes with k-spectrum H, the k-spectral segments of the obtained k-segmentation of H occur with different possible orders that are specific to each genome.

If we change perspective and assume to start from the genomic sequence itself, we may compute its spectral segmentation in a more efficient way. We used this procedure in order to compute the k-segmentations of real genomes [19].

Algorithm 4.2 (computing a k-segmentation from a genome)
A k-mer $\alpha \in D_k(\mathbb{G})$ is *univocally elongated* in (the k-spectrum of) G if $|\{\beta \in D_k(G) : \alpha[2...k] = \beta[1...k-1]\}| = 1$. A Boolean array A, such that $|A| = |\mathbb{G}|$, is initialized to be *false* in every position.

For each $\alpha \in D_k(G)$, such that α is *univocally elongated* in $\mathbb{G}$, the algorithm sets as $true$ all the positions of A of the set $pos(\alpha, \mathbb{G}) = \{i : \mathbb{G}[i, ..., i+k-1] = \alpha\}$. Then, the k-segmentation is retrieved by scanning for consecutive runs of $true$ assignments in A. A k-spectral segment is a substring $\mathbb{G}[i...j]$ of $\mathbb{G}$ such that $A[l] = true\ \forall l : i \leq l \leq j$. □

All the notions based on genome spectra can be extended with respect to a dictionary D, by replacing $spec_k(\mathbb{G})$ by $spec_D(\mathbb{G})$, the multiset of strings of D with their occurrence multiplicities in $\mathbb{G}$:

$$spec_D(\mathbb{G}) = \{(\alpha, mult_{\mathbb{G}}(\alpha)) \mid \alpha \in D\}.$$

Of course:

$$spec_k(\mathbb{G}) = spec_{D_k(\mathbb{G})}(\mathbb{G}).$$

We call k-**spectral ambiguity** of $\mathbb{G}$ the number of different genomes having the same spectrum of $\mathbb{G}$. A more practical index of ambiguity, we call k-**permutation ambiguity** is the average number of k-spectral segments that are spectrally k-permutable.

We define the **absolute spectrum** of a genome $\mathbb{G}$, denoted by $spec_{Abs}(\mathbb{G})$, as the spectrum of $\mathbb{G}$ relative to its absolute dictionary $Abs(\mathbb{G})$.

Spectra, when visualized by means of heat maps, highlight the graphical similarity of the human chromosomes, and their dissimilarity with the mitochondrial chromosomes, as it is shown in Fig. 4.12. The figure shows 6-mers spectrum of each

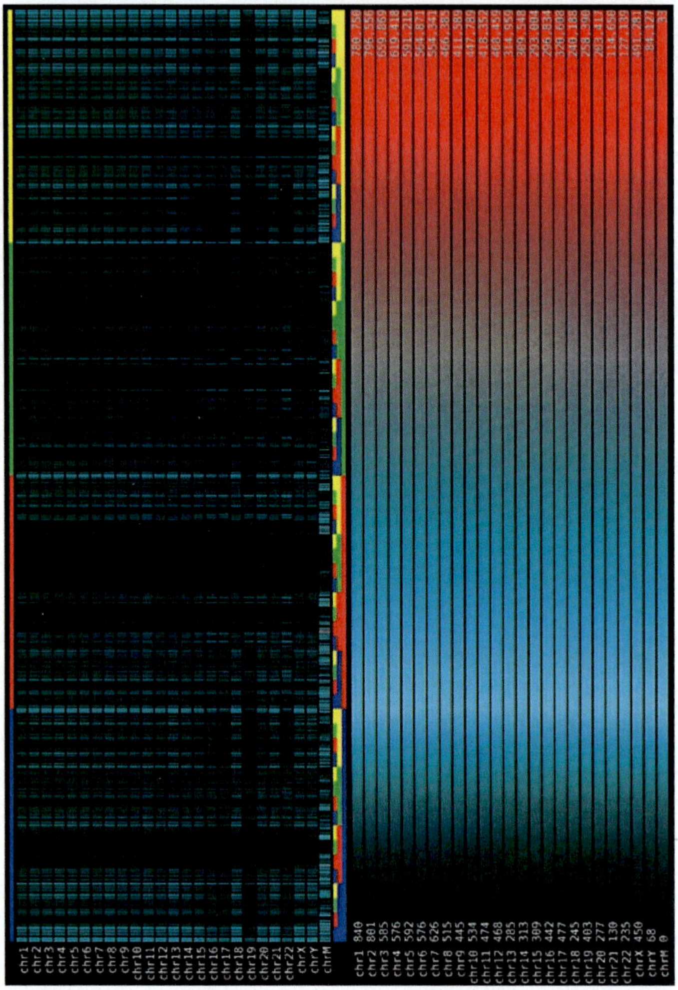

Fig. 4.12 6-mer multiplicity heat map of all human chromosomes. The bottom four colors ruler indicates the 3-mer prefix of each column block. The original image was created using 4096 (equals to 4^6) pixels

chromosome by mapping multiplicities to a given palette. The correspondence to the black colour is performed by mapping such colour to the minimum multiplicity of each chromosome, and the maximum multiplicity is mapped to the red colour.

4.7 Elongation and Segmentation

Elongation and segmentation are two crucial aspects of genomes.

A segmentation of a genome is a multiset of strings covering the genome, where each string is equipped with its multiplicity of occurring in the genome.

We have already met elongation and segmentation in the analysis of genome spectra.

If we consider that genomes are products of a long evolutive process where longer and longer sequences were assembled from very short fragments, then we easily realize how much elongation and segmentation are important in selecting strings that compose genomes.

For value of $k \geq 20$ the set of all possible strings, become gigantic, and those really occurring in genomes are only an infinitesimal part. This means that after this value elongation of k-mers is a very selective operation.

A **right elongation** of a non-empty string is a string of which α is a prefix, while a **left elongation** is a string of which it is a suffix. String α is **right elongated** by β in a genome $\mathbb{G}$ if α is prefix of β that occurs in $\mathbb{G}$. Moreover α is **right univocally elongated** by β if there is no string γ different from β such that α is right elongated by γ in $\mathbb{G}$. Finally, α is **right maximally elongated** by β in $\mathbb{G}$ if α is right elongated by β but β is not univocally right elongated by any string in $\mathbb{G}$. Analogous definitions can be given for left elongation.

Figure 4.13 is the representation of an elongation tree. Starting from a small number of "seeds" an entire genome can be expressed by showing all the elongations of the seeds covering the given genome. Figures 4.16, 4.15 and 4.14 are examples of representations of small genomes that are based on elongation trees (using circles and colors). What is surprising in these representations is the strong symmetry of the

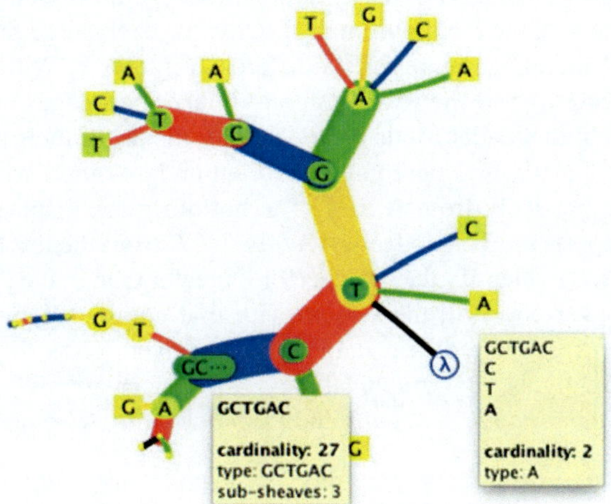

Fig. 4.13 An elongation tree (from [20])

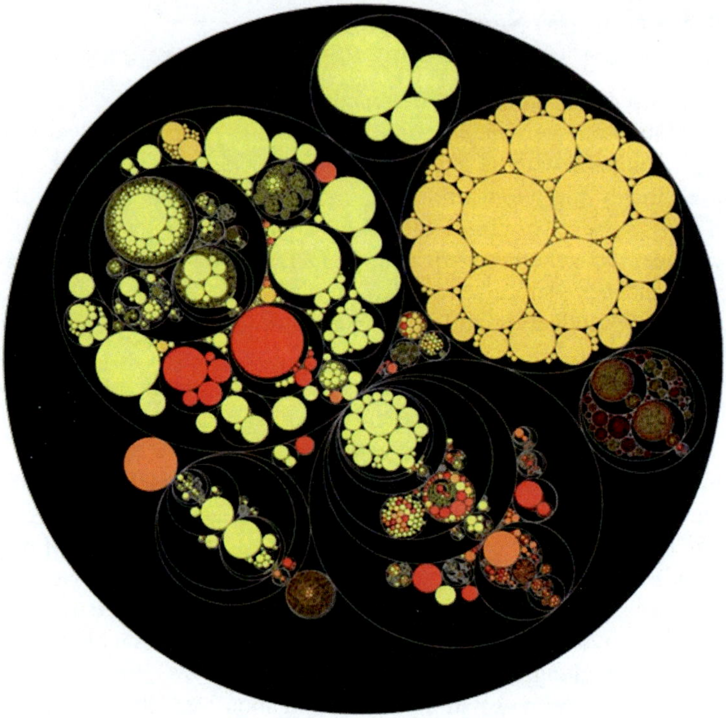

Fig. 4.14 An elongation tree represented by means of circles (from [20])

structures and the self-similarity between parts. Similarly, Figs. 4.17, 4.18 and 4.19 show the multiplicity of k-mers for k from 1 to 6 mapped to the gradients that are shown in the bottom part of the image. A multiplicity value of 0 is mapped to black, instead, the maximum multiplicity, for each value of k, is mapped to red. Each row of the image represents a different value of k. Within each row, k-mers are sorted according to the lexicographic order. The images also contain coloured rules that help the reader to identify k-mer by means of their prefix. The top ruler reports the 4 1-mers, lexicographically sorted and coloured with blue, red, green and yellow, respectively from A to T. The bottom ruler, displaced at the centre of the image, reports the 3-mers from AAA to TTT from the left to the right. Such a ruler can be used to identify the k-mers (for k greater than 3, the 256 columns rising from a given 3-mer cover all the k-mers with that common 3-prefix). Although, the resolution of the reported heat maps is poor, nevertheless they give an overall view, which highlights, in a graphical way, differences among species, such as *Homo sapiens, Caenorhabditis elegans* and *Caenorhabditis elegans*.

Proposition 4.19 *Hapaxes of a genome are, right and left, univocally elongated in the genome.*

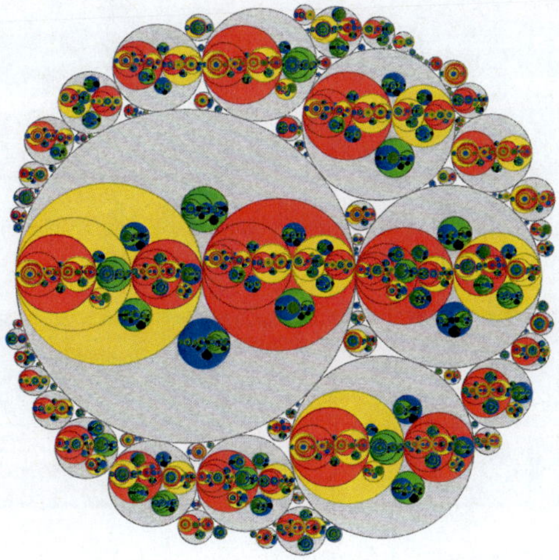

Fig. 4.15 Genome of Coliphage Phi-X174 represented by 103 seeds of maximum elongation length 15 (from [20])

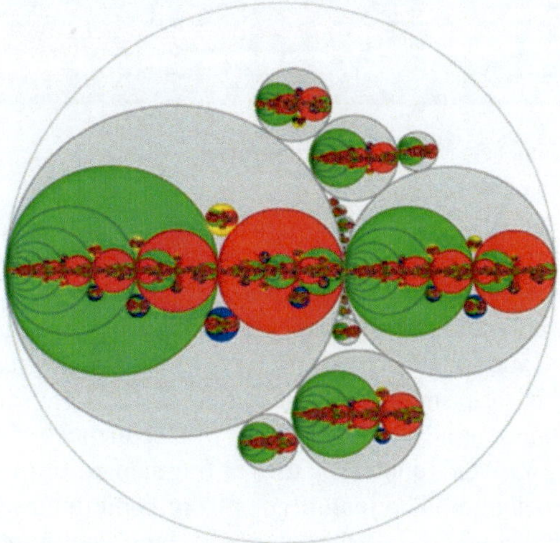

Fig. 4.16 Genome of Carsonella Ruddii (from [20])

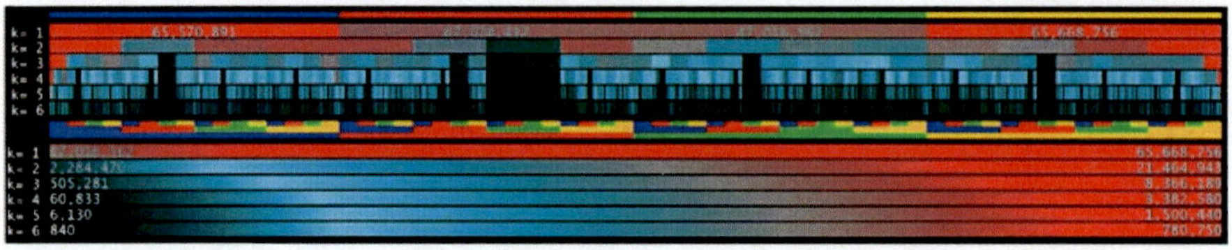

Fig. 4.17 Word branching heat map of the human chromosome 1. Values of k are $(1, 2, 3, 4, 5, 6)$ and displayed in the same order, from top to bottom side

Fig. 4.18 Word branching heat map of *Escherichia coli*

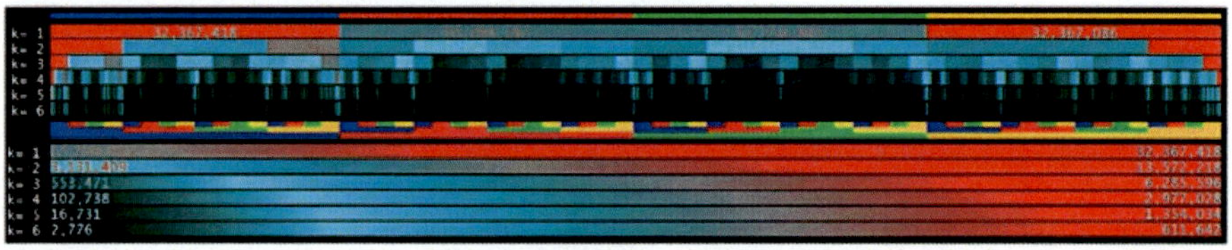

Fig. 4.19 Word branching heat map of *Caenorhabditis elegans*

Genome Sequencing consists of the correct identification of the nucleotide sequence of a given genome. The methods used for sequencing are based on many different biochemistry and biophysics techniques (everywhere changing and improving their efficiency, cost, and reliability, according to technical progress in nanotechnology). However, the basic principle of sequencing remains the same in all different technologies, which essentially are algorithms of a sequence reconstruction from a dictionary. Namely, a given genome is always fragmented in pieces, called **reads**, smaller than a given length. The length goes from some tenths to some thousand bases, according to the specific methodologies, which continuously increase both length and efficiency. These fragments are read by means of mechanisms of biochemical of physical nature. A sequencing technique deduces the most probable sequence containing all the reads. Usually, reads need to be amplified, by providing many copies of each one. Crucial factors of the process are not only the average length of reads but also their redundancy, that is, the average number of reads associated with any genome position. It is experimentally proved that the shorter reads are, the greater their redundancy has to be, in order to obtain good sequencing. Reads are assembled, that is, they are concatenated with overlaps, and the longer and more precise are the overlaps, the more, their concatenations are reliable. By applying iteratively, and hierarchically, the process of overlap concatenation of longer substrings of the original genome is generated, up to a unique sequence that corresponds to the putative genome from which the initial reads were generated (probability and precision can be quantified by using reasonable hypotheses about the errors introduced in the assembly process).

Two main kinds of sequencing can be distinguished: "de novo sequencing" and re-sequencing. In the second case, the genome sequence is already known for a genome of the same species, therefore reads are aligned, or mapped, (with some possible mismatches) to a whole sequence of reference. On the contrary, in the first case, this reference is missing, therefore the greater number of possible solutions have to be considered, evaluated and discriminated.

Here we want to point out an aspect that, in many cases, could be a methodological limitation of the sequencing approach. In fact, since the first approaches to human sequencing project, the genome to which the process above outlined was applied did not correspond to the genome of a single individual but was a sort of mixing derived from a small population of individual genomes. This choice was motivated by the purpose of obtaining, in such a way, something more general and representative of the biological species of selected individuals.

In our opinion, now we are in a position where the new technological efficiency allows us to reverse this kind of approach. Namely, if a big number of different individual genomes are sequenced, from the analysis of their differences and of their common structure, we can deduce, in a more correct way, what can be considered the essence of all genomes of a given species. This new perspective could provide new ideas and new applications in many fields, from medical science to a better understanding of evolutionary mechanisms.

The following proposition gives an interesting result that characterizes the organization of genomes in terms of particular kinds of spectra.

Given a genome $\mathbb{G}$ and a k-mer α and $x \in \{A, T, C, G\}$, we denote by

$$e_{\mathbb{G},x}(\alpha)$$

the number of times α is right elongated with $x \in \mathbb{G}$. Of course, it holds that:

$$mult(\alpha) = e_{\mathbb{G},A}(\alpha) + e_{\mathbb{G},T}(\alpha) + e_{\mathbb{G},C}(\alpha) + e_{\mathbb{G},G}(\alpha).$$

From this equation, it easily follow that when two genomes have the same $(n + 1)$-spectrum, then, necessarily, they have also the same n-spectrum. *Vice versa*, from the knowledge of $e_{\mathbb{G},x}(\alpha)$ for $x \in \mathbb{G}$, it is easy to compute the $(n + 1)$-spectrum of $\mathbb{G}$ from its n-spectrum.

We denote by $fel(\mathbb{G})$ the *fully elongated length* of $\mathbb{G}$, that is the length k such that $e_{\mathbb{G},x}(\alpha) > 0$ for any $x \in \{A, T, C, G\}$, and for any possible k-mers α.

Proposition 4.20 *Let $k = mgl(\mathbb{G}) + 1$ (the minimal hapax greatest length incremented by 1), then if a genome $\mathbb{G}$ has the same k-spectrum of $\mathbb{G}_1$, both genomes $\mathbb{G}$ and $\mathbb{G}_1$ have the same minimal hapaxes that occur in both genome in the sane* *relative order. Moreover, the two genomes may differ for possible swaps of strings located between their common minimal hapaxes.*

Proof First, having the two genomes the same $(n+1)$-spectrum the multiplicity of any n-mer α is the same in both genomes.

Now, if $\mathbb{G}$ and $\mathbb{G}_1$ have the same k-spectrum, then they have the same stings of length k with multiplicity 1. But for the definition of k, all the minimal hapaxes of $\mathbb{G}$ are substrings of the k-mers with multiplicity 1, and moreover, the two genomes have the same spectra for values smaller than k, therefore, they have the same minimal hapaxes. Moreover, in both genomes hapaxes of length k, including minimal hapaxes, are not k-permutable, because two of them cannot have the same $(k-1)$-prefix and $(k-1)$-suffix, being some of these strings hapaxes too.

In conclusion, repeats of a genome have to occur in the regions comprised between (non k-permutable) consecutive k-hapaxes. In particular, being hapaxes elongation of minimal hapaxes, repeats surely occur between consecutive minimal hapaxes. □

Given a genome $\mathbb{G}$, for any k, we denote by

$$card_k(\mathbb{G})$$

the number or different k-mers that occur in $\mathbb{G}$. The value:

$$\log_2 card_{mgl}(\mathbb{G})$$

is the number of bits necessary for representing all the minimal hapaxes of $\mathbb{G}$, while $2mgl(\mathbb{G})$ is the number of bits necessary for representing all possible mgl-mers. Therefore, the following fraction is inversely proportional to the "informational selectivity" of $\mathbb{G}$, because the smaller it is, the greater the information that the genome selects:

$$\frac{\log_2 card_{mgl}(\mathbb{G})}{2mgl(\mathbb{G})}.$$

This is only a partial indication, because does not count the selection related to the repeats between minimal hapaxes (see later on).

Let $D_{mh}(\mathbb{G})$ be the dictionary of the minimal hapaxes of $\mathbb{G}$. For any $\alpha \in D_{mh}(\mathbb{G})$, we denote by $e_{\mathbb{G}}(\alpha)$ the unique right elongation of $\alpha \in \mathbb{G}$, that does not include the minimal hapax following α in $\mathbb{G}$, then the set:

$$E_{mh} = \{e_{\mathbb{G}}(\alpha) | \alpha \in D_{mh}(\mathbb{G})\}$$

determines a segmentation of $\mathbb{G}$ where all the segments of E_{mh} are hapaxes.

The set $D_{mhr}(\mathbb{G})$ of maximal repeats occurring between two consecutive minimal hapaxes of $\mathbb{G}$ constitutes a dictionary such that the whole genome $\mathbb{G}$ is a sequence having the following form (h_i is the minimal hapax of occurrence order i):

$$\mathbb{G} = h'_1 R_1 h'_2 R_2 \ldots h'_m R_m$$

where, for $i \leq m$, h'_i is a hapax including (possibly non strictly) the minimal hapax h_i, while $R_i \in D_{mhr}(\mathbb{G})$, and for different values of $i < j < m$ it may be $R_i = R_j$. The same form holds for all genomes having the same $(mgl + 1)$-spectrum as $\mathbb{G}$, apart from a possible ordering of the inter-hapax repeats.

4.8 Genome Informational Laws

Random Genomes are generated by a Bernoulli random process extracting at each step one of the letters in Γ, having each letter the same probability of being extracted. In a next chapter we will study such kinds of genomes, and we will show that the empirical k-entropies of random genomes of length n, for any $k \geq 1$, are bounded by $2\log_2 n = 2LG$. Moreover, entropy $E_{2LG}(\mathbb{G})$, for any genome $\mathbb{G}$, is bounded by the value of E_{2LG} in random genomes of the same length as $\mathbb{G}$, therefore the following inequalities hold:

First Law	$LG(\mathbb{G}) < E_{2LG}(\mathbb{G}) < 2LG(\mathbb{G})$

By recalling the meaning of entropy of an information source, as mean information of the data emitted by the source, the relation above expresses a very general property of genomes of a given length, saying that for k-mers of binary logarithmic length, the mean information quantity emitted by a genome, as an information source of k.mers, is between the logarithmic length and the double logarithmic length of the genome. This fact has been verified for a wide class of genomes of different types [4].

Given a genome $\mathbb{G}$, two values are associated to it: the **entropic component** $EC(\mathbb{G})$ and the **anti-entropic component** $AC(\mathbb{G})$, defined by the following equations:

$$EC(\mathbb{G}) = E_{2LG}(\mathbb{G}) - LG(\mathbb{G})$$

$$AC(\mathbb{G}) = 2LG(\mathbb{G}) - E_{2LG}(\mathbb{G})$$

shortly denoted by EC and AC, when the genome is understood. From these definitions follows that:

$$LG = AC + EC.$$

The **anti-entropic fraction** $AF(\mathbb{G})$ is given by:

$$AF(\mathbb{G}) = AC(\mathbb{G})/LG(\mathbb{G})$$

while the entropic fraction $EF(\mathbb{G})$ is given by:

$$EH(\mathbb{G}) = [EC(\mathbb{G}) - AC(\mathbb{G})]/LG$$

and from the definitions above it follows that $EH = 1 - 2AF$.

Let LX be the index $LX_k = |D_k(G)|/|G|$ with $k = 2LG(\mathbb{G})$ (D_k the dictionary of k-mers). In [4] some informational laws have been identified and expressed, in terms of the entropic indexes introduced above, which were tested for 70 genomes of different types (from bacteria to primates). Here we add to the previously given law, a second law relating the entropic components to the lexical index LX, associated to the length LG:

Second law	$EC > LX \cdot AC$

Apart from the specific contents of these informational laws, it is worthwhile to notice the interest of such kinds of laws for a deep comprehension of the informational nature of genomes.

A third law, given in the next section, will put in evidence the role of the two entropic components of genomes in expressing their evolutionary complexity.

4.9 Genome Complexity

Let $\mathbb{G}$ a genome of length n. A complexity measure is defined in [4], which is based on the entropy and anti-entropy of genomes defined in the previous section. For genomes of 70 different biological species such a measure agrees in a very surprising manner with the evolutionary complexity of species, in the usual phylogenetic sense. The entropic component is responsible for the chaotic component pushing genomes to change, while the anti-entropic component refers to the structure and to biological functions of genomes.

The index BB, called also **Bio-Bit**, is given by:

$$BB = \sqrt{LG}\sqrt{AF}(1 - AF)^3$$

it corresponds to the measure of evolutionary complexity. Its definition, apart from the factor $\sqrt{LG}$, is a sort of logistic map $x(1-x)$, where factors are weighted by exponents: $x^{\frac{1}{2}}(1-x)^3$, according to experimental evaluations over a great number of genomes, at different levels in the evolutionary scale.

The equilibrium between the two entropic components expresses an essential aspect of evolution, as a consequence of deep abstract informational dynamics that is intrinsic to genome texts. The complexity measure that we define increases in time by disclosing a tendency opposite to that one driven by the second principle of thermodynamics. In fact, physical entropy concerns the individual genomes, as physical molecular structures. Species genomes, which are abstractions of a purely symbolic nature, increase their complexity as texts that in time exploit richer forms of life. In this perspective, death is a necessity of life, which refers only to the physical entities supporting individual instances of purely symbolic texts. In this perspective, the expression "purely symbolic text" is somewhat paradoxical, because intends something like an "unwritten text". In other words, the written texts are of physical

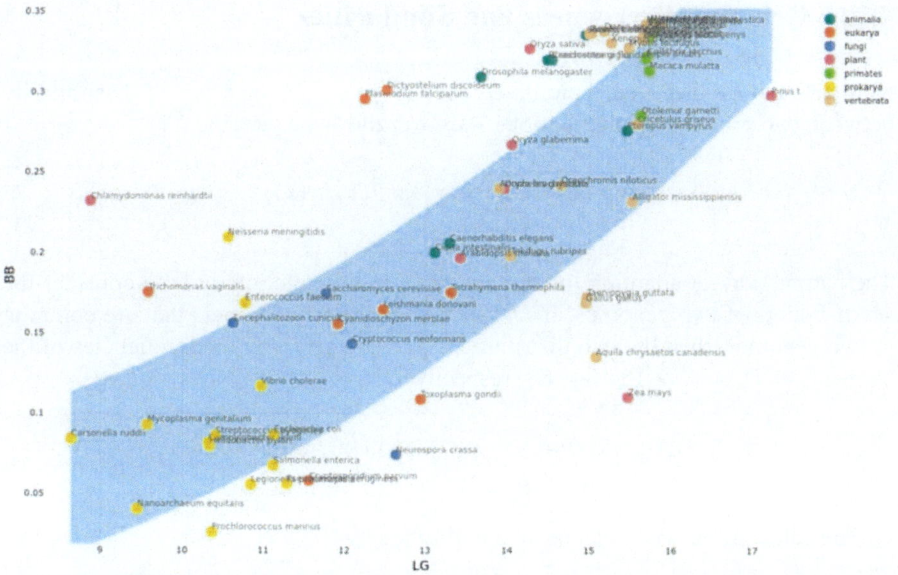

Fig. 4.20 Genome Biobit ordering. In the abscissa the logarithmic length and in the ordinate the biobit value

Fig. 4.21 Biobit ordering. The up-right corner of Fig. 4.20 enlarged

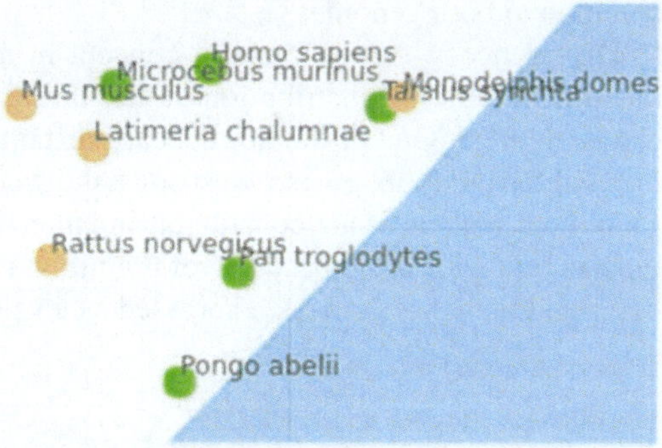

nature, subjected to perish, while what is common to a set of such texts is their form, which is independent from any individual physical text. This abstract form is what supports life development along with evolution.

The last law given in [4] is here formulated, as a third informational law on genomes. This law claims that $BB(\mathbb{G})$ expresses the genomic complexity of genome $\mathbb{G}$ according to the following genomic complexity ordering $\prec$ (Figs. 4.20 and 4.21).

Third Law	$BB(\mathbb{G}_1) < BB(\mathbb{G}_2) \Rightarrow \mathbb{G}_1 \prec \mathbb{G}_2$

4.10 Genome Divergences and Similarities

Kullback and Leibler's entropic divergence $D(p, q)$ is a measure of dissimilarity between two probability distributions p, q over the same variable X:

$$D(p, q) = \sum_{x \in \widehat{X}} p(x) \lg(p(x)/q(x)).$$

The natural way of adapting this notion to a genomic dictionary is to consider the set of k-mers of two genomes. If $D_k(\mathbb{G}_1, \mathbb{G}_2)$ is the set of k-mers that are common to two genomes, then the two distributions p, q are given by the frequencies of the elements of $D_k(\mathbb{G}_1, \mathbb{G}_2)$ in $\mathbb{G}_1, \mathbb{G}_2$, respectively:

$$\Delta_k(\mathbb{G}_1, \mathbb{G}_2) = \sum_{x \in D_k(\mathbb{G}_1, \mathbb{G}_2)} p(x) \lg(p(x)/q(x))$$

The value $\Delta_k(\mathbb{G}_1, \mathbb{G}_2)$ can be also multiplied by

$$|\Gamma_k|/|D_k(\mathbb{G}_1, \mathbb{G}_2)|$$

for considering the inaccuracy due to missing frequencies of k-mers that are not common to both genomes.

Divergence is surely a useful concept in measuring the disagreement between random variables or distributions. But other measures of similarity or dissimilarity can be defined. One of them is Jaccard distance that is based on a dictionary. In its original formulation, given two strings, the dictionaries of k-mers (for some value of k) for both sequences are considered and then the ratio between the cardinality of the intersection with the cardinality of the union is considered as a degree of similarity between the two sequences, expressed as a value between 0 and 1:

$$J_k(\alpha, \beta) = \frac{|D_k(\alpha) \cap D_k(\beta)|}{|D_k(\alpha) \cup D_k(\beta)|}$$

A variant of this definition takes into account also the multiplicities of k-mers, by considering the union of two dictionaries and taking as a similarity measure the ratio between a numerator, given by the sum of the minimum of the multiplicities of k-mers in the union dictionary, and a denominator, given by the maximum of the multiplicities of k-mers in the union dictionary.

$$J_k'(\alpha, \beta) = \frac{\sum_{w \in D_k(\alpha) \cup D_k(\beta)} min\{mult_\alpha(w), mult_\beta(w)\}}{\sum_{w \in D_k(\alpha) \cup D_k(\beta)} max\{mult_\alpha(w), mult_\beta(w)\}}$$

Jaccard distance is $1 - J_k(\alpha, \beta)$, which can also be defined by considering the symmetric difference of k-mer dictionaries:

$$\frac{|D_k(\alpha) \cup D_k(\beta)| - |D_k(\alpha) \cap D_k(\beta)|}{|D_k(\alpha) \cup D_k(\beta)|}.$$

Another way of considering similarity is given by **attribute vectors**. An attribute vector is a sequence of real values giving a measure to any attribute of a list of n attributes. This means that any string α is associated with a real vector of $V(\alpha) \in \mathbb{R}^n$. For example, if we chose n k-mers as attributes, the corresponding vector for a string consists of the frequencies that the chosen k-mers have in the string. In this case the similarity between two strings α, β is given by the cosine of the angle between their vectors $V(\alpha), V(\beta)$, where $< \cdot | \cdot >$ represents the usual scalar product in $\mathbb{R}^n$:

$$\frac{< V(\alpha) | V(\beta) >}{< V(\alpha) | V(\alpha) >< V(\beta) | V(\beta) >}.$$

Distance is a classical concept of mathematics, and a similarity index is a sort of opposite notion because expresses a vicinity with respect to some parameters of interest. If distances D and similarities S are considered to range in the $[0, 1]$ interval, then $S = 1 - D$.

The generalization of Euclidean distance is the Minkowski distance d_k of parameter $k \in \mathbb{N}$ and $k > 0$ defined between vectors of $\mathbb{R}^n$:

$$d_k(X, Y) = \sqrt[k]{\sum_{1 \leq i \leq n} |X(i) - Y(i)|^k}.$$

For $k = 2$ we obtain the usual Euclidean distance, while for $k = 1$ we obtain the sum of distances in the different cartesian dimensions of the space (this distance is also called Manhattan distance).

Minkowski distances can be also normalized in several ways by dividing the terms by suitable values, depending on the kind of application (average values of standard deviation within some population of vectors).

Cebychev distance between two vectors X, Y is the maximum among the absolute value of differences $|X(i) - Y(i)|$.

A distance, frequently used between binary sequences, is the Hamming distance, which is a special case of Manhattan distance between two boolean vectors (of the same length), that is, the number of positions where they have different values:

$$h(X, Y) = \sum_{1 \leq i \leq n} |X(i) - Y(i)|.$$

Another kind of dissimilarity for genomic applications is obtained by considering the dictionary of k-mers, for some value k. then if $\mu_1(w), \mu_2(w)$ are the multiplicities of the k-mer w, in two genomes $\mathbb{G}_1, \mathbb{G}_2$, then when we set $\mu_{1,2}(w) = |\mu_1(w) - \mu_2(w)|$ and a dissimilarity measure given by:

$$\sum_{w \in \Gamma^k} \mu_{1,2}(w).$$

If $\mu_0 = max\{\mu_{1,2}(w) \mid w \in \Gamma^k\}$, then the following sum is a normalized way for expressing the multiplicity differences of k-mers between two genomes:

$$\sum_{w \in \Gamma^k} \frac{\mu_{1,2}(w)}{\mu_0}.$$

The same approach can be extended by replacing multiplicities with probabilities and by normalizing with the sums of probability differences:

$$p_{1,2}(w) = |p_1(w) - p_2(w)|$$

$$p_* = \sum_{w \in \Gamma^k} p_{1,2}(w)$$

$$\sum_{w \in \Gamma^k} \frac{p_{1,2}(w)}{p_*}.$$

The notions of distance, proximity, similarity, dissimilarity, and divergences, focusing on different specific aspects, are of particular importance in the analyses of genomes and have a crucial role in the comparisons of genomes, and in the identification of common structures and evolutionary or functional differences.

Let us say that two discrete variables are **joint** when they assume values that we can arrange in two sequences where values in the same position are considered each other corresponding. In this case, the **covariance** between two variables is defined by:

$$cov(X, Y) = \frac{1}{n}\sum_{i=1}^{n}(x_i - E[X])(y_i - E[Y]) = E[(X - E[X])(Y - E[Y])]$$

where $E(X)$, $E(Y)$ are the mean expected values of random variables X, Y. Pearson's **correlation** is defined by means of covariance by:

$$P_{X,Y} = \frac{cov(X, Y)}{\sigma_X \sigma_Y}$$

where σ_X and σ_Y are the **standard deviations** of X, Y respectively, that is. the root square of the corresponding variances σ_X^2 and σ_Y^2, that is, by denoting also by μ the mean value $E[X]$:

$$\sigma_X^2 = E[(X - E[X])^2] = \sqrt{\frac{\sum_{i=1}^{n}(x_i - \mu)^2}{n}}$$

When $P_{X,Y} > 0$ variables X, Y are positively correlated, while when $P_{X,Y} < 0$, variables X, Y are negatively correlated, or anti-correlated. Another way of measuring the degree of agreement, between two discrete random variables, is the following Kendal ratio (after Maurice Kendall, a statistician who introduced it in 1938):

$$\frac{C(X, Y) - D(X, Y)}{\frac{n(n-1)}{2}}$$

where $C(X, Y)$ is the number of values u, v of the two variables X, Y such that $\frac{u-E[X]}{v-E[Y]} > 0$, while $D(X, Y)$ is the number of values u, v such that $\frac{u-E[X]}{v-E[Y]} < 0$, and $\frac{n(n-1)}{2}$ is the number of all possible pairs of values assumed by the two variables.

Correlation indexes are very often useful in random variables analysis and can be defined in many different ways according to the kinds of comparisons more appropriate to the contexts of interest.

Spearman correlation is defined by ordering the values of two variables X, Y and by comparing values according to pairs: $(x_1, y_1), (x_2, y_2), \ldots, (x_n, y_n)$. Then, the values of two variables are ordered in **ranks**. In other words, the values of the first variable are ordered from the smallest value to the greatest one, and also the second variable values are ordered in the same manner (for simplicity it is assumed that values do not repeat). The rank of a value is its position in these orderings. The rank differences of two corresponding pairs are summed giving a correlation between the two variables (possibly normalized with the standard deviations of the rank variables). Other versions of this correlation count the numbers of rank inversions between contiguous values of the first variable and between contiguous pairs of the second variable, by comparing these numbers in some way.

Other indexes, from economics or statistics, could be relevant also for many kinds of sequences. In fact, when a real variable provides values related to the individuals of a population, it is interesting to analyze how much these values differ or agree, on average. The origin of this kind of analysis was concerned, for example, with the richness distribution in a society, or with the water abundance in the rivers of the world. As the Italian economist Wilfred Pareto showed, the most typical situations follow the so-called 20/80 rule: the 20% of the population has 80% of all resources of the whole population. If we plot the correspondences between the two percentages, that is, for each population percentage the corresponding percentage of corresponding resources, then we get a curve and its integral area gives a measure of the fairness level of the resource distribution among individuals. In this context, Gini's coefficient (after the Italian statistician Corrado Gini, at beginning of '900) is defined (for all possible pairs) as:

$$g = \frac{\sum_{i \neq j} |x_i - x_j|}{n(n-1)}$$

possibly, by normalizing this value with the average of all values.

Any notion of information is naturally related to a kind "distance", in a wide sense. Information is uncertainty, that is, a sort of difference w. r. t. the certainty of necessary events, mutual information is a divergence between two particular probability distributions, and in general, meaningfulness can be considered a sort of distance from randomness. Therefore, this discussion highlights the crucial role of "general" notions of distance between probability distributions. The attribute "general" wants to stress the possibility that some properties of classical metric distances can be relaxed. A necessary property is the positivity of distance and its null value for pairs on the diagonal, but symmetry and transitivity are generally too strong. In fact, in some non-homogeneous spaces the distance from two points can differ in the two verses (the top of a hill and its basis are two positions with asymmetric distances), and for the same reason, the distance along two edges of a triangle can be shorter that the direct distance between two extremities of one edge.

A genomic distance directly involving entropy is the following that we call Distance Entropy. Given two genomes G, G', of the same length that are segmented in m disjoint and adjacent parts, called segments: $G_1, G_2, \dots G_m$ and $G'_1, G'_2, \dots G'_m$, and given a distance d between strings, if d_i is the value of d between segments at position i, and p_i the probability of occurring of the value d_i (within a prefixed approximation range), then the Distance Entropy between the two genomes is given by:

$$DE(G, G') = -\sum_{1}^{m} p_i \lg p_i.$$

A different notion of information strictly based on random variables is Fisher's information. It expresses the role that a parameter θ has in determining the probability distribution of a given random variable (depending on θ).

Technically, the likelihood of a probability distribution $p(X, \theta)$ is a sort "inverse probability", that is, a function giving the probability of observing $X = x$ for a given value of the parameter θ; in terms of conditional probability we can write:

$$p(X, \theta) = prob(X|\theta).$$

Fisher information associated with a random variable X is the variance of the logarithmic derivative (w.r.t. θ), also called score, of $p(X, \theta)$ computed for the true parameter θ_0, that is, the value of θ corresponding to the probability distribution p_X of X. The mean score of $p(X, \theta)$ at θ_0 is zero. In fact, let us denote it by $E_{\theta_0} p(x, \theta)$, abbreviated as E_θ:

$$E_\theta = \int_{-\infty}^{+\infty} \left(\frac{\partial}{\partial \theta} \ln p(x, \theta) \right) p(x, \theta_0) dx = \int_{-\infty}^{+\infty} \frac{p'(x, \theta)}{p(x, \theta)} p(x, \theta_0) dx$$

whence

$$E_{\theta_0} = \int_{-\infty}^{+\infty} \frac{p'(x, \theta_0)}{p(x, \theta_0)} p(x, \theta_0) dx$$

that is:

$$E_{\theta_0} = \left[\frac{\partial}{\partial \theta} \int_{-\infty}^{+\infty} p(x, \theta) dx \right]_{\theta_0} = 0.$$

Therefore, being $E_{\theta_0} = 0$, the variance of $p(X, \theta)$ corresponds to:

$$F(p(X, \theta)) = \left(\frac{\partial p}{\partial \theta} \ln p(X, \theta) \right)^2_{\theta_0}.$$

We do not give more details on Fisher's information, but we want to remark that in both notions of information, due to Shannon and Fisher respectively, information is related to probability, and in both cases, the logarithm is used in the functional dependence from probability.

Finally, an issue related to the significance of information is the notion of p-value. In a very wide sense, the p-value associated with a datum is the inverse of its informational relevance, that is a value expressing its level of casualty. In a technical sense, this value corresponds to the probability that the datum was generated by chance/error in the process of generating a series of observed values. In more precise terms, given a random variable X, the p-value of $x_0 \in \widehat{X}$ is the probability of observing a value equal or more extremal than x_0, if the observed value was generated according to a random variable Z that is assumed as reference for casualty. In this formulation "more extremal" can be specified in many ways, depending on the shape of Z: (left-tail, right-tail, or doubly-tailed). The most natural choice of Z is a normally distributed variable (casual errors distribute normally) with mean and standard deviation evaluated from the elements of $\widehat{X}$. But other choices can be determined by specific features of the given variable X.

A threshold α (usually 5% or 2%) is associated with a p-value $p(x_0)$ for x_0, such that, if $p(x_0) < \alpha$, then the value is considered significant, otherwise x_0 is considered as due to chance.

4.11 Lexicographic Ordering

Lexicographic ordering is an essential and intrinsic characteristic of strings. Assume an ordering $<$ over the symbols of the alphabet. This order is extended to all the strings over the alphabet by the following conditions. Given two strings α, β we set:

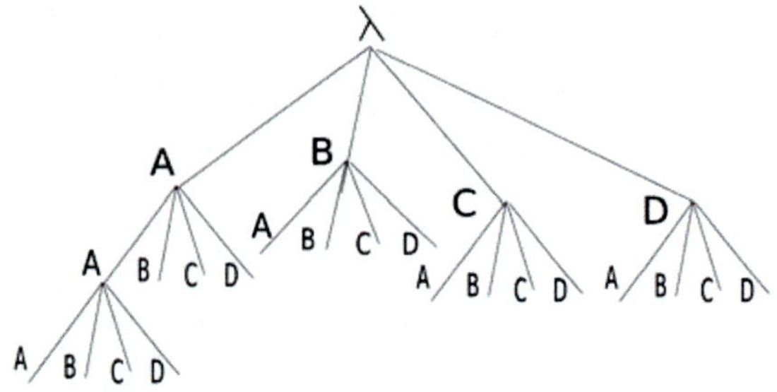

Fig. 4.22 A tree of a lexicographic order of a quaternary alphabet. Any path is a string. Ordering is left-right and top-down. Two levels are full and the first four strings of the third level are given

(i) $\alpha < \beta$ if $|\alpha| < |\beta|$
(ii) if $|\alpha| = |\beta|$ let i be the first position from the left where α and β differ, then $\alpha < \beta$ if $\alpha[i] < \beta[i]$.

It is easy to represent the lexicographic ordering by a tree where the empty string is put as a label on the root and a son node is added, from the left to the right, to the root for every symbol of the alphabet, which is associated as a label of the node. Then, iteratively the same method is applied to every node added at the last level, in such a way that every node identifies the string concatenating all the labels of the path from the root to the node. In this way, a string α precedes lexicographically another string β if the node to which α is associated is at the previous level or on the left of the node to which β is associated (Fig. 4.22).

The general rule of generating strings in the lexicographic order follows directly from the structure of the tree given above. First, the symbols of the alphabet are generated in their order (consider four symbols) in consecutive orders and periods:
A, B, C, D (First Order)
$A(A, B, C, D), B(A, B, C, D), C(A, B, C, D), D(A, B, C, D) =$
$AA, AB, AC, AD, BA, BB, BC, BD, CA, CB, CC, CD, DA, DB, DC,$
DD (First Period)
In general, if:
$\alpha, \beta, \ldots \eta$
are the string of the period of n (in their lexicographic order), then:
$A\alpha, A\beta, \ldots A\eta, B\alpha, B\beta, \ldots B\eta, C\alpha, C\beta, \ldots C\eta, D\alpha, D\beta, \ldots D\eta$
are the strings of period $n + 1$.
Essentially, this generation schema is that one, based on **orders** and **periods**, introduced by the great Greek mathematician Archimedes of Syracuse (213–212 BC) [10, 21], for generating big numbers. Starting from k digits $A_1, A_2 \ldots A_k$ periods are generated iteratively. First period has k orders:
$A_1, A_2, \ldots, A_k$
$A_1A_1, A_1A_2, \ldots, A_1A_k$
$A_2A_1, A_2A_2, \ldots, A_2A_k$

....

$A_k A_1, A_k A_2, \ldots, A_k A_k$

In general, from a period P the next period is obtained by producing an order for each element $x \in P$, in such a way that if $x < y$ all the elements of the order produced by x precede all the elements of the order produced by y.

The interpretation of Archimedes was multiplicative, this means that using decimal symbols the expression 375 is intended to denote 5×10^{37} (Archimedes defined exponential numbers, by using geometric progressions, where the final value of an order coincides with the ratio of the following order). Whereas, as we are going to show, the number expressing the position of 375 in the lexicographic ordering (over 10 symbols) corresponds to an additive interpretation of digits, analogous to that one of the usual decimal system, which, differently from the lexicographic representation, includes a digit for denoting zero [21–23].

Lexicographic ordering can easily be defined by induction. Let's assume a finite ordered alphabet (for example the usual letters). Let α, β be any two strings over the alphabet and x, y any two elements of this alphabet, then:

$$\lambda < a < b < \cdots < z$$

$$\alpha < \alpha x$$

$$\alpha < \beta \Rightarrow \alpha x < \beta y$$

$$\alpha < \beta \Rightarrow x\alpha < x\beta.$$

(this definition is correct and complete).

Lemma 4.1 (Lexicographic recurrence) *Let S be an (ordered) finite alphabet of k symbols where, for any $x \in S$ a number $\rho_S(x)$ is assigned, with $1 \leq \rho_S(x) \leq k$, then the lexicographic enumeration number $\rho_S(\alpha x)$, $\alpha \in S^*$, $x \in S$ satisfies the following equation:*

$$\rho(\alpha x) = k\rho(\alpha) + \rho(x). \tag{4.1}$$

Proof Let us identify S with $\{1, 2, \ldots k\}$, and let us consider the lexicographic enumeration k-tree having $\{1, 2, \ldots k\}$ ordered (from left) sons of the root, then $\{11, 12, \ldots 1k\}$ as ordered sons of 1 and so on for 2, 3, ... The tree grows by iterating the same procedure at each level, with k ordered sons $\{\alpha 1, \alpha 2, \ldots \alpha k\}$ for any node α. Now, if $\beta \prec \alpha$ denotes that β precedes lexicographically α, then it easy to realize that ($|\ |$ denotes the cardinality of the set, the string λ is included):

$$\rho(\alpha) = |\{\beta \in S^* \mid \beta \prec \alpha\}|$$

therefore, according to the way lexicographic ordering is defined, being $|S| = k$, the strings preceding αx consist of two disjoint sets:

(1) $\{\beta y \in S^* \mid \beta \prec \alpha, y \in S\} \cup \{\lambda\}$

(2) $\{\alpha y \in S^* \mid y \in S, y \prec x\}$

The first set is obtained by the set $\{\beta \in S^* \mid \beta \prec \alpha\}$ of $\rho(\alpha)$ elements elongated with all possible symbols of S. These strings are of course $k\rho(\alpha)$, but to them the string λ has to be added, giving in the overall $k\rho(\alpha) + 1$ elements. The second set has $\rho(x) - 1$ elements, therefore, being disjoint, the number of strings preceding αx are:

$$\rho(\alpha x) = k\rho(\alpha) + 1 + \rho(x) - 1. \tag{4.2}$$

according to the formula asserted by the lemma.

Proposition 4.21 (The lexicographic Theorem) *Let S be an (ordered) finite alphabet of k symbols where, for any $x \in S$ a number $\rho_S(x)$ is assigned, with $1 \leq \rho_S(x) \leq k$, then the lexicographic enumeration number $\rho_S(\alpha)$ of string $\alpha \in S^*$ is given by (for empty string λ, $\rho(\lambda) = 0$).*

$$\rho(\alpha) = \sum_{i=1}^{|\alpha|} \rho(\alpha[i])k^{|\alpha|-i} \tag{4.3}$$

Proof By iterating Eq. 4.2 we get Eq. 4.3 asserted by the theorem. Let $\alpha = a_1a_2a_3 \ldots a_n$:

$$\rho(\alpha) = k\rho(a_1a_2a_3 \ldots a_{n-1}) + \rho(a_n) =$$

$$k[k\rho(a_1a_2a_3 \ldots a_{n-2}) + \rho(a_{n-1})] + \rho(a_n) =$$

$$k^2\rho(a_1a_2a_3 \ldots a_{n-2}) + k\rho(a_{n-1}) + \rho(a_n) =$$

$$k^2[k\rho(a_1a_2a_3 \ldots a_{n-3}) + \rho(a_{n-2})] + k\rho(a_{n-1})]\rho(a_n) =$$

$$k^3\rho(a_1a_2a_3 \ldots a_{n-3}) + k^2\rho(a_{n-2}) + k\rho(a_{n-1})]\rho(a_n).$$

Therefore, going on up to the end, when ρ is applied only to symbols $a_1, a_2, a_3, \ldots a_n$, we get exactly what the theorem asserts. □

Corollary 4.1 *Given a number $k > 0$, any non-null natural number n, such that $k^{m-1} \leq n \leq k^m$, is univocally representable by a linear combination of powers k^i with all coefficients belonging to $\{1, 2, \ldots k\}$:*

$$n = \sum_{i=1}^{m} c_i k^{m-i}.$$

We remark that the equation above is the same expressing the number representation with respect to a base $k > 1$, with the difference that in the case of usual positional representation digit 0 is allowed, and for all the coefficients c_i the following inequalities hold:

Table 4.1 Lexicographic sum

+	1	2	3	4
1	2	3	4	11
2	3	4	11	12
3	4	11	12	13
4	11	12	13	14

Table 4.2 Lexicographic difference, where * means decreasing one unit of the leftmost digit

−	1	2	3	4
1	4*	1	2	3
2	3*	4*	1	2
3	2*	3*	4*	1
4	1*	2*	3*	4*

Table 4.3 Lexicographic product

×	1	2	3	4
1	1	2	3	4
2	2	4	12	14
3	3	12	21	24
4	4	12	24	34

Table 4.4 Lexicographic division, where * means one-decreasing of the leftmost digit and subscript denotes the remainder

/	1	2	3	4
1	1	2	3	4
2	2_1*	1	1_1	1_2
3	1_2*	2	2_1*	2_2*
4	1_1*	1_2*	1_3*	2

$$0 \leq c_i < k.$$

Arithmetic operations for the lexicographic number representation [10] are based on the operation tables expressed by symbols 1, 2, 3, 4. It is important to remark that this number representation is zeroless, but at the same time it is positional [24] (Tables 4.1, 4.2, 4.3 and 4.4).

4.12 Suffix Arrays

A problem occurring very often with strings is the search for a given substring, that is, answering the question:

$$\beta \in sub(\alpha).$$

Suffix trees were introduced in years 1970, to efficiently solve the question above and a great number of other related problems. The main intuition behind the notion of a suffix tree is very simple. In fact, any substring of a given string is a prefix of some suffix. Then, given a string w let us consider the set suff(w) of its suffixes and aggregate iteratively them by means of common prefixes. Suffixes are easily defined by induction: suff(λ) = $\{\lambda\}$ and suff(αx) = suff(α) $\cup$ $\{\alpha x\}$. Let us consider a root λ with the elements of suff(w) labelling edges exiting from the root. If $\gamma\alpha \in$ suff(w) and $\gamma\beta \in$ suff(w), then we replace the edge with an edge labelled by γ and connected to two edges labelled by α and β respectively.

This rule is applied again when other common prefixes are found between two strings. It was proven that the suffix tree can be constructed in linear time (with respect to the length of w). However, the main result of the suffix tree data structure is that the substring problem can be solved in a time that is linear with respect to the length of the substring, independently from the length of the superstring where it is searched. The following figure shows a suffix tree, where an extra symbol \$ was used in order to have all the suffixes as leaves. From a suffix tree, it is possible to deduce the occurrence multiplicity of any substring and the positions where it occurs. We do not give further details on the suffix tree, preferring to focus on suffix arrays (Fig. 4.23).

Suffix arrays [25] are structures derived from the suffix tree. The suffix array of a given string α is based on the sequence of all suffixes of the string ordered according to their alphabetic ordering, that is, the lexicographic order where the conditions about the string length are removed: a string α precedes a string β when in the first position j where they differ α has a symbol preceding the symbol at j position of β. The main intuition of the suffix array is the fact that any substring is a prefix of some suffix of the string. Hence, a suffix array is expressed by two vectors:

(i) vector SA_α such that $SA_\alpha(i)$ is the position of α where the suffix of alphabetic order i occurs in α;

Fig. 4.23 The suffix tree of BANANA

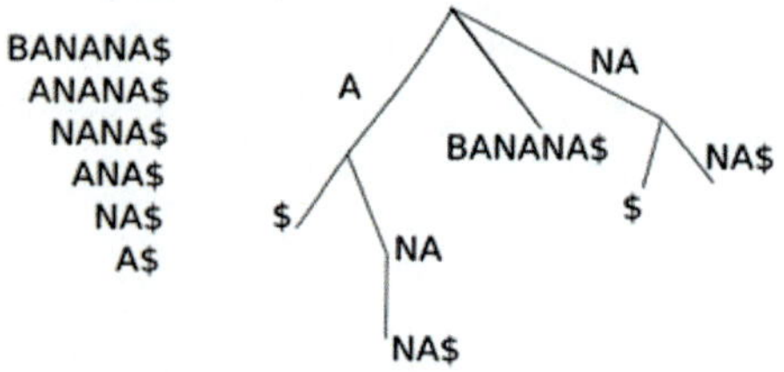

Table 4.5 Suffix array of BANANA

Index	Suffix	Position	LCP
1	A	6	0
2	ANA	4	1
3	ANANA	2	3
4	BANANA	1	0
5	NA	5	0
6	NANA	3	2

(ii) vector LCP_α such that $LCP_\alpha(i)$ is the length of the longest common prefix between suffix $SA_\alpha[1, i]$ and suffix $SA_\alpha[1, i-1]$ ($LCP_\alpha[1, 1]$ is set to 0).

A k-interval of LCP_α is a contiguous range $[i, j]$ of indexes of vector LCP_α such that:
$LCP_\alpha[i] < k$
$LCP_\alpha[l] \geq k$, for $i < l \leq j$

In other words, k has to be less than the value of LCP_α in correspondence to the left index of the interval, but cannot be less than LCP_α on the other indexes of the interval.

Suffixes belonging to the same k-interval, for some positive integer k, share the same k-prefix, also called the prefix of the k-interval (Table 4.5).
In the suffix array above interval $[1, 3]$ is a 1-interval, and interval $[2, 3]$ is a 3-interval. Analogously, interval $[5, 6]$ is a 2-interval. Of course, any k-interval is a h-interval for any $h < k$. What is important is that all the suffixes of a k-interval share their k-prefixes. Therefore if α is the k-prefix of the first suffix of an interval, then it is the prefix of all the suffixes of that interval.

The computational cost of realization of such structures is linear with respect to the length of the string to which are applied. However, when they are constructed the substring search becomes logarithmic with respect to the length of the overall string.

Software IGTools [8] is entirely based on a suitable extension of suffix arrays and makes possible the efficient computation of many genomic indexes based on very big genomic dictionaries that would be intractable without a so powerful data structure.

References

1. Manca, V., Scollo, G.: Explaining DNA structure. Theor. Comput. Sci. **894**, 152–171 (2021)
2. Bonnici, V., Manca, V.: An informational test for random finite strings. Entropy **20**(12), 934 (2018)
3. Manca, V.: The principles of informational genomics. Theor. Comput. Sci. (2017)
4. Bonnici, V., Manca, V.: Informational laws of genome structures. Sci. Rep. **6**, 28840 (2016). http://www.nature.com/articles/srep28840. Updated in February 2023

5. Manca, V.: Information theory in genome analysis. In: Membrane Computing, LNCS 9504, pp. 3–18. Springer, Berlin (2016)
6. Manca, V.: Infogenomics: genomes as information sources. Chapter 21, pp. 317–324. Elsevier, Morgan Kauffman (2016)
7. Bonnici, V., Manca, V.: Recurrence distance distributions in computational genomics. Am. J. Bioinform. Comput. Biol. **3**, 5–23 (2015)
8. Bonnici, V., Manca, V.: Infogenomics tools: a computational suite for informational analysis of genomes. J. Bioinform. Proteomics Rev. **1**, 8–14 (2015)
9. Manca, V.: On the logic and geometry of bilinear forms. Fundamenta Informaticae **64**, 261–273 (2005)
10. Manca, V.: On the Lexicographic Representation of Numbers, pp. 1–15. Cornell University Library (2015). ArXiv.org
11. Manca, V.: Infobiotics: Information in Biotic Systems. Springer, Berlin (2013)
12. Manca, V.: From biopolymer duplication to membrane duplication and beyond. J. Membr. Comput. (2019). https://doi.org/10.1007/s41965-019-00018-x
13. Annunziato, A.: DNA packaging: nucleosomes and chromatin. Nat. Educ. **1**(1), 26 (2008)
14. Lynch, M.: The Origin of Genome Architecture. Sinauer Associate. Inc., Publisher (2007)
15. Manca, V.: A marvelous accident: the birth of life. J. Proteomics Bioinform. **11**, 135–137 (2018)
16. Franco, G.: Biomolecular Computing—Combinatorial Algorithms and Laboratory Experiments. Ph.D. Thesis, Dipartimento di Informatica Università di Verona (2006)
17. Manca, V., Franco, G.: Computing by polymerase chain reaction. Math. Biosci. **211**, 282–298 (2008)
18. Castellini, A., Franco, G., Manca, V.: A dictionary based informational genome analysis. BMC Genomics **13**, 485 (2012)
19. Bonnici, V., Franco, G., Manca, V.: Spectrality in genome informational analysis. Theor. Comput. Sci. (2020)
20. Lombardo, R.: UnconventionalComputations and Genome Representations, Ph.D. Thesis, Dipartimento di Informatica Università di Verona (2013)
21. Manca, V.: Lezioni Archimedee. Quaderni della Biblioteca Alagoniana, Siracusa, N. 6 (2020)
22. Manca, V.: Il Paradiso di Cantor. La costruzione del linguaggio matematico, Edizioni Nuova Cultura (2022)
23. Manca, V.: Arithmoi. Racconto di numeri e concetti matematici fondamentali (to appear)
24. Boute, R.T.: Zeroless positional number representation and string ordering. Am. Math. Mon. **107**(5), 437–444 (2000)
25. Abouelhoda, M.I., Kurtz, S., Ohlebusch, E.: Replacing suffix trees with enhanced suffix arrays. J. Discret. Algorithms **2**, 53–86 (2004)

Chapter 5
Information and Randomness

Probability, Randomness, Information

We have already seen that information is intrinsically related to probability, analogously randomness is the basis of probability. In fact, we will show as any random distribution can be obtained from a purely random sequence of values. In this chapter, the mutual relationships between the notions of probability, information, and randomness will be considered in more details, by emphasizing some important aspects that directly link information with randomness.

5.1 Topics in Probability Theory

In this section, we will analyze some important aspects of probability theory that help to understand subtle points of crucial probabilistic notions.

Monty Hall's Problem

A famous debate about probability is the *three doors quiz* also known as the *Monty Hall problem*, by the name of the TV conductor in a popular American show. There are three doors A, B, and C. A treasure is behind only one of them, and you are asked to bet on the lucky door. Let's assume that the player chooses the door C, but another chance is given to him/her, and the conductor opens the door B showing that the treasure is not there. The player is free of confirming the preceding choice of C or change by betting on A. The question is: *What is the more rational choice?*

Many famous mathematicians concluded that there is no reason for changing (Paul Erdös, the number one of 20th mathematicians, was among them). However,

V. Manca and V. Bonnici, *Infogenomics*, Emergence, Complexity and Computation 48,
https://doi.org/10.1007/978-3-031-44501-9_5

by using Bayes' theorem it can be shown that changing the door is the better choice. This fact was even confirmed by computer simulations of the game.

Proposition 5.1 *In the three doors quiz, after that, the player bets on C and the conductor opens the door B showing that the treasure is not behind B, the more rational choice of the player is changing the choice betting on A.*

Proof Let denote by D the event: The player chooses C and after that the door B is opened (showing no treasure behind B).

Let denote by A_T the event: The treasure is behind the door A.

Let us apply Bayes Rule for computing $P(A_T|D)$:

$$P(A_T|D) = P(D|A_T)P(A_T)/P(D)$$

$P(D|A_T) = 1$, because given A_T, after the player choice of C, the conductor can open only B without stopping the game. Of course $P(A_T) = 1/3$. Moreover, $P(D) = 1/2$, because after the choice of C two alternative cases are possible; if the treasure is behind A (with probability $1/3$), the conductor can open only B, while if the treasure is behind C (with probability $1/3$) the conductor can indifferently open A or B, therefore the probability of opening B is $1/3 + 1/6 = 1/2$. In conclusion, by substituting the values of probabilities in the formula above we obtain:

$$P(A_T|D) = (1 \times 1/3)/1/2 = 2/3.$$

Therefore, after player's choice of C, when the conductor opens B the probability $P(A_T)$ changes from its previous value of $1/3$ to the new value of $2/3$. □

The solution to the Monty-Hall problem given above tells us something of very general about the power of Bayes' approach to probability. People sustaining that the more rational behaviour is maintaining the initial bet support their position by observing that the world has not changed by the opening of door B. In fact, this is true. But what results from our proof is that even if the world is unchanged, our information about the world changes when the door is open (and no treasure is found). We can make more evident the increase of information after this event if we generalize the quiz to 100 doors. Let's assume that the player guesses that the treasure is behind the second door. Moreover, let us assume that the conductor opens doors 1, 3, 4, 5, ... 69, 71, 72, ...100 (all doors apart 2 and 70). In end, all the doors, apart from doors 2 and 70, are open showing no treasure. Is it more rational to maintain the initial bet, or to change by betting on door 70? In this case, is evident that if door 70 is closed, while door 2 maintains the initial probability of having the treasure (the conductor cannot open it), on the contrary, door 70 cumulates all the information of the events opening doors different from 70.

In other words, information affects probability and changes our knowledge of the world. But the knowledge that only our rational investigation of the world can obtain does not coincide with reality. No scientific theory tells us what reality exactly is, but can only increase our knowledge about reality. In this sense, Bayes' theorem intercepts one of the crucial points about the essence of science, by giving new evidence that probability and information concern with the scientific attitude that in Latin is expressed by "Modus conjectandi" ("as we judge that the things are", the title of Bernoulli's treatise was *Ars Conjectandi*), which distinguishes from *Modus Essendi* (as the things are). We can only construct reasonable (and approximate) models about reality, but no definitive and absolute certainty can be reached by any scientific investigation. For this reason, science, since the 19th century, is based on probability and its progress is strongly based on probabilistic models. Moreover, these models are not due to temporary states of ignorance, but as modern physics tells us, they are intrinsically related to the essence of the scientific statements about reality (see the section *Information in Physics* of this chapter).

de Moivre-Laplace Theorem

Given an urn having white balls with percentage p ($1-p$ the percentage of black balls), the probability $p(n,k)$ of having k withe balls in extracting n balls is:

$$p(n,k) = \binom{n}{k} p^k q^{n-k}. \tag{5.1}$$

The mean value of k, in the extractions of n balls, is given by np because it is the mean of the sum of n independent boolean random variable X (1 white and 0 black) each having the mean $E(X) = 1 \times p + 0 \times (1-p) = p$. Analogously, the standard deviation can be evaluated. In fact, the square deviation of any boolean random variable X is ($q = 1-p$):

$$(1-p)^2 p + (0-p)^2 q = q^2 p + p^2 q = qp(q+p) = qp$$

therefore the sum of n independent variable X has square deviation npq, and standard deviation $\sqrt{npq}$.

Let us denote by $a_{n,k}$ the probability of having $n+k$ white balls when $2n$ balls are extracted in an urn containing the same number of white and black balls. In the following, we make explicit in $\simeq$ the variable with respect to which the asymptotic equivalence holds by writing $\simeq_n$.

Proposition 5.2 (De Moivre-Laplace Theorem) $a_{n,k} \simeq_n \frac{1}{\sqrt{\pi n}} e^{-k^2/n}$.

Proof According to Bernoulli's binomial distribution:

$$a_{n,k} = (2n)_{(n+k)!}/(n+k)!2^{-2n}$$

which can be rewritten as:

$$a_{n,k} = [(2n)_{n!}(n)_{k!}/(n+k)!]\,2^{-2n} = 2^{-2n}\frac{(2n)!}{n!n!}\frac{n!(n)_{k!}}{(n+k)!}.$$

therefore $a_{n,k}$ is the product of two terms; a first term:

$$2^{-2n}\frac{(2n)!}{n!n!} \qquad (I)$$

and a second term:

$$\frac{(n)_{k!}n!}{(n+k)!} \qquad (II).$$

Using Stirling approximation:

$$n! \simeq_n \tfrac{n^n}{e^n}\sqrt{2\pi n}$$

for the term (I) we have:

$$2^{-2n}\tfrac{(2n)!}{n!n!} \simeq 2^{-2n}\tfrac{(2n)^{2n}}{e^{2n}}\tfrac{e^n}{n^n}\tfrac{e^n}{n^n}\frac{\sqrt{2\pi 2n}}{\sqrt{2\pi n}\sqrt{2\pi n}}$$

$$\simeq 2^{-2n}\tfrac{2^{2n}n^{2n}}{e^{2n}}\tfrac{e^n}{n^n}\tfrac{e^n}{n^n}\frac{\sqrt{2\pi 2n}}{\sqrt{2\pi n}\sqrt{2\pi n}} \simeq \frac{\sqrt{2\pi n}\sqrt{2}}{\sqrt{2\pi n}\sqrt{2\pi n}} \simeq \frac{1}{\sqrt{\pi n}}.$$

Let us rewrite the second term (II):

$$\frac{(n)_{k!}n!}{(n+k)!} = \frac{n}{n+1}\frac{n-1}{n+2}\frac{n-2}{n+3}\cdots\frac{n-k+1}{n+k}$$

which can again be rewritten as:

$$\left(1-\frac{1}{n+1}\right)\left(1-\frac{3}{n+2}\right)\left(1-\frac{5}{n+3}\right)\ldots =$$

$$\left(1-\frac{1}{n+1}\right)^{\frac{n+1}{n+1}}\left(1-\frac{3}{n+2}\right)^{\frac{n+2}{n+2}}\left(1-\frac{5}{n+3}\right)^{\frac{n+3}{n+3}}\ldots \simeq_n$$

$$e^{-\frac{1}{n+1}}e^{-\frac{3}{n+2}}e^{-\frac{5}{n+3}}\ldots \simeq_n$$

$$e^{-(1+3+5\ldots)/n} \simeq_n$$

$$e^{-k^2/n}.$$

In conclusion we have:

$$a_{n,k} \simeq_n \frac{1}{\sqrt{\pi n}} e^{-k^2/n} \tag{5.2}$$

□

Now, let us consider the following probability distribution (the integral of $e^{-\frac{1}{2}\frac{x}{\sigma}^2}$ over the whole real line is well-known to be $\sigma\sqrt{2\pi}$):

$$\frac{1}{\sigma\sqrt{2\pi}} e^{-\frac{1}{2}(\frac{x}{\sigma})^2}$$

and replace x and σ with the deviation from the mean μ and with the standard deviation, respectively, of the Bernoulli variable considered above, which has mean n (n white balls and n black balls), deviation from the mean $n + k - n = k$, and standard deviation $\sqrt{2n/4} = \sqrt{n/2}$:

$$\frac{1}{\sqrt{2\pi n/2}} e^{-\frac{1}{2}(\frac{k}{\sqrt{n/2}})^2} = \frac{1}{\sqrt{\pi n}} e^{-k^2/n}$$

which coincides with the equation of the De Moivre-Laplace theorem. This means that this theorem is a particular case of a more general distribution, as we will show in the next section.

Normal, Poisson and Exponential Distributions

De Moivre-Laplace theorem can be extended to the case of a probability $p(n, k)$ of k successes in n extractions with p and q the percentage of white and black balls, by giving the following equation:

$$p(n, k) = \frac{1}{\sqrt{npq}\sqrt{2\pi}} \; e^{-\left(\frac{1}{2}\left[\frac{(k-np)}{\sqrt{npq}}\right]^2\right)}$$

where np is the mean value and $\sqrt{npq}$ the mean standard deviation. If these two values are replaced by μ and σ, respectively, we obtain the general Normal or Gaussian distribution $n_{\mu,\sigma}$ of mean μ and standard deviation σ:

$$n_{\mu,\sigma}(x) = \frac{1}{\sigma\sqrt{2\pi}} \; e^{-\left(\frac{1}{2}\left[\frac{(x-\mu)}{\sigma}\right]^2\right)}.$$

This law can be found in many random phenomena, going from the distribution of the height in a population of individuals to the distribution of the errors in the measures of a physical quantity.

Poisson law can be derived from Bernoulli's law under suitable hypotheses, typical of rare events: a very big number n of possible events and a probability p very small of having k successful events. In this case, if we call λ the product np, the probability is derived with the following reasoning. Let us replace p by $\frac{\lambda}{n}$ in Bernoulli formula of urn extraction, then the probability $P_{n,k}$ of k successes (white ball) in n extractions is:

$$P_{n,k} = \binom{n}{k}\left(\frac{\lambda}{n}\right)^k\left(1-\frac{\lambda}{n}\right)^{n-k} \tag{5.3}$$

$$P_{n,k} \simeq \binom{n}{k}\left(\frac{\lambda}{n}\right)^k e^{-\lambda} \tag{5.4}$$

$$P_{n,k} \simeq \frac{n_{k!}}{k!}\frac{\lambda^k}{n^k}e^{-\lambda} \tag{5.5}$$

$$P_{n,k} \simeq \frac{n_{k!}}{n^k}\frac{\lambda^k}{k!}e^{-\lambda} \tag{5.6}$$

$$P_{n.k} \simeq \frac{\lambda^k}{k!}e^{-\lambda} \tag{5.7}$$

The right member of Eq. (5.7) is called Poisson distribution, after the mathematician who introduced it.

Usually, a time real variable t is introduced in such a way that $\lambda = np$ is the mean of successes in the unit time, whereas λt is the mean of successes in the time interval t, thus the Poisson distribution $P_\lambda(k)$ (of parameter λ) is set by:

$$P_\lambda(k) = e^{-\lambda t}\frac{(\lambda t)^k}{k!} \tag{5.8}$$

If we set $k = 0$ in the Poisson distribution (5.8) we obtain:

$$P_\lambda(0) = e^{-\lambda t} \tag{5.9}$$

therefore $1 - P_\lambda(0)$ is the probability of a success at end of the t interval, that is, t is the waiting time for the first success. However, this is a continuous distribution (t is real), thus we consider the density distribution of the waiting time in a process following Poisson distribution, by taking the derivative of $1 - P_\lambda(0) = 1 - e^{-\lambda t}$. In conclusion, this waiting time is given by the exponential distribution (of parameter λ):

$$E_\lambda(t) = \lambda e^{-\lambda t}. \tag{5.10}$$

The **geometric distribution** of probability p gives the probability of a success extraction exactly at step k. It is the discrete version of exponential:

$$(1-p)^{k-1}p.$$

The **hypergeometric distribution** refers to urn extractions of balls without inserting the extracted balls. What is the probability of having k successes in n extractions? If N is the total number of balls, and m the number of white balls (success balls), then the probability is given by:

$$\binom{m}{k}\binom{N-m}{n-k}/\binom{N}{n}$$

where the denominator counts all the possible ways of extracting n balls, while the numerator counts the ways of choosing k white balls among the all m white balls multiplied by the number of ways of choosing the remaining $n-k$ balls among all the remaining $N-m$ black balls.

The **negative binomial**, also called Pascal's distribution, gives the probability of waiting for n extractions for having k successes. Let p be the probability of success and q the probability of failure. Of course, when the kth success occurs exactly at extraction n we had $n-1$ extractions with $k-1$ successes, which have a probability $\binom{n-1}{k}p^{k-1}q^{n-k}$. If we multiply this value by the probability of having success with the following extraction we obtain:

$$\binom{n-1}{k}p^k q^{n-k}.$$

The probability that after k extraction other $m \geq 0$ extractions need for k successes results, with easy calculations, to be (this is the reason for the term "negative binomial"):

$$\binom{-k}{m}p^k(-q)^m.$$

Multinomial distribution refers to the probability of extracting n balls of k types where n_i are of type i and p_i are the corresponding probabilities for $i = 1, 2, \ldots, k$:

$$\binom{n}{n_1, n_2, \ldots, n_k}p^{n_1}p^{n_2}\ldots p^{n_k}.$$

Laws of large numbers

The first law of large numbers was found by Jakob Bernoulli and can be informally expressed by saying that when an event with probability p occurs within a set of n events, then the number m of times it occurs tends to pn as n increases, or also $p \simeq m/n$. Bernoulli was so impressed by this result that he named it "Theorema Aureus". A deep meaning of this asymptotic behaviour is the intrinsic link between time and probability. In time, events occur with a multiplicity corresponding to their probabilities. In other words, when we know their probabilities, then we can reasonably predict their occurrence, and vice versa, their probabilities can be reasonably inferred from their occurrences in time. In the following, we present a stronger form of Bernoulli's golden theorem, which is due to Cantelli (further generalized by Emil Borel in the context of Measure Theory).

The following Lemma puts in evidence the main property used in the proof of strong law of large numbers.

Lemma 5.1 (Cantelli 1917) *Let $(E_k | k \in \mathbb{N})$ be a sequence of events such that $P(E_k) = a_k$ and $\sum_{k=0}^{\infty} a_k \in \mathbb{R}$, then with probability 1 only a finite number of events E_k can occur.*

Proof From the hypothesis $\sum_{k=0}^{\infty} \in \mathbb{R}$ we have that $\forall \varepsilon > 0$ there exists an $n > 0$ such that;

$$\sum_{k>n}^{\infty} a_k < \varepsilon.$$

If we assume that infinite events E_k can occur with a probability $\delta > 0$, then, by the equation above, such a δ has to be less than any real value, therefore δ cannot be different from 0. Whence, with probability 1 only a finite number of events can occur. □

From the previous lemma, the following proposition follows.

Proposition 5.3 (Strong Law of Large Numbers) *Given a Bernoulli boolean sequence where a success 1 occurs with probability p, a failure 0 with probability q, and S_n the sum of sequence up to n, then for every positive real ε, the following inequality*

$$\left| \frac{S_n}{n} - p \right| > \varepsilon \tag{5.11}$$

with probability 1 holds only a finite number of times.

(Feller [1])

Random walks

A random walk (in a discrete plane) can be represented as a sequence in $\{0, 1\}$ by encoding by 0 a horizontal step and by 1 a vertical step. An equivalent way of representing a random walk is a walk ± 1, which is a sequence in $\{1, -1\}$ (diagonal, anti-diagonal instead of horizontal, vertical). In this case, starting from the origin of the plane $(0, 0)$ the sum of steps without sign gives, at any time, the value of the ordinate of the current position, whereas the algebraic sum of steps gives the abscissa of the current position. Therefore, if 1 steps are p, while -1 steps are q, respectively, then $r = p - q$, and $n = p + q$ and are the coordinates of the position reached at end of steps ± 1. Of course, $(n + r)/2 = (p + q - q + p)/2 = p$. The number of walks ± 1 from the origin to (n, r), with p steps 1 and q steps -1 is given by:

$$N_{n,r} = \binom{p+q}{p} = \binom{n}{(n+r)/2}. \tag{5.12}$$

Reflexion Principle (André) Let w be a walk ± 1 from the point $A = (x, y)$ to the point B of the discrete plane. Let us assume that this walk touches or crosses the abscissa in a point $C = (c, 0)$. Then, a w' walks ± 1 going from the point $(A' = (x, -y)$ to B is uniquely associated to w, where all the steps of w' before reaching the point C pass along the symmetric points of walk w, whereas from C to B w and w' coincide. □

The following proposition is a consequence of the Reflexion Principle.

Proposition 5.4 (Bertrand) *The number $N_{n,r}$ of walks ± 1 from $(0, 0)$ to (n, r) that, apart from the first position, do not touch or cross the abscissa is $\frac{r}{n} N_{n,r}$.*

The Ballot Problem (Bertrand) Given a ballot between two competitors A, B who at the end cumulate p and q votes, respectively $(p > q)$. What is the probability that the winner has at any time of the ballot a greater number of votes?

The solution to the problem is obtained by applying Bertrand's Theorem. In fact, any counting of votes is a random walk ± 1 with p steps 1 and q steps -1. All these walks are $N_{n,r}$ $(n = p + q, r = p - q)$. Moreover, ballots, where the winner is always in advantage, correspond to walks ± 1 that do not touch or cross the abscissa, whence, according to Bertrand's Theorem, their number is: $(p - q)/(p + q)N_{n,r}$. Therefore, the ratio between always advantageous ballots and all the possible ballots is:

$$(p - q)/(p + q)N_{n,r}/N_{n,r} = (p - q)/(p + q).$$

which is the probability of having always advantageous ballots.

Random processes and Arcsine Law

A random process is a family of random variables:

$$\{\chi_i\}_{i \in I}$$

therefore, random values are generated in correspondence to the values i in the set I. Usually i corresponds to time (real number), which can be discrete if I is restricted to sets of natural or integer numbers.

Therefore, in a random process not only the probability of taking values is considered, but the probability of taking values at a given time. A random walk is a random process, and a process of Bernoulli extractions is a random process too. In both cases, at each step, a value of a Boolean random variable is generated. In this sense, a genome can be seen as a random process at several levels, the level of single letters generation, or the levels of string generation (for all possible segmentations of the genome).

Given a random process, we can define many random variables associated with the process, that depends on the random variables underlying the process. However, even when the process is purely random, as the deMoivre-Laplace theorem shows, nevertheless such derived random variables are not purely random but follow specific probability distributions, such as the normal, Poisson, or exponential distributions. In this sense, these distributions are often called **laws of chance**. Let us give some examples based on a process of Bernoulli extractions.

1. the probability that (exactly/at least/at most) k successes occur in n steps;
2. the probability that (exactly/at least/at most) k failures occur between two successes;
3. the probability of (exactly/at least/at most) k advantages of one type are cumulated in n steps;
4. the probability that the first parity (abscissa touching) occurs at step $2n$;
5. the probability that the last parity (abscissa touching) occurs at step $2n$;
6. the probability of (exactly/at least/at most) k consecutive successes/failures (runs) occur in $2n$ step;
7. the probability that (exactly/at least/at most) k successes occur when exactly n failures occur (Pascal or negative binomial distribution).

Let us consider a particular case, that is, a process of Bernoulli extractions with success probability 1/2. In $2n$ extractions, for the law of large numbers, when all $2n$ extractions are concluded, the probability of having a parity, with n successes and n failures, is surely very close to 1. But, for $k < n$ what is the probability that the last situation of parity happens at extraction $2k$?

The previous question is equivalent to evaluating, in a random walk, the number of paths touching the abscissa at step $2k$ that do not touch again this axis up to step $2n$ (included). This number is given by:

$$\alpha_{2k,2n} = u_{2k}u_{2n-2k}$$

where, by Stirling approximation:

$$u_{2n} = \binom{2n}{n} 2^{-2n} \simeq 1/\sqrt{\pi n}.$$

Therefore,

$$\alpha_{2k,2n} = u_{2n-2k} = \frac{1}{\sqrt{\pi k}} \frac{1}{\sqrt{\pi(n-k)}} = \frac{1}{\pi\sqrt{k(n-k)}}.$$

Then if we set:

$$x_k = \frac{k}{n}$$

and

$$f(x_k) = \frac{1}{\pi\sqrt{x_k(1-x_k)}}$$

we obtain:

$$\alpha_{2k,2n} = \frac{f(x_k)}{n}.$$

Now, we recall that the derivative of $\arcsin(x)$ function is given by:

$$\arcsin'(x) = \frac{1}{\sqrt{1-x^2}}$$

whence:

$$\frac{2}{\pi} \arcsin'(\sqrt{x}) = \frac{1}{\pi\sqrt{x(1-x)}} = f(x)$$

therefore, for $0 < x < 1$:

$$\sum_{k=1}^{nx} \alpha_{2k,2n} \approx \int_0^x f(x)dx = \frac{2}{\pi} \arcsin(\sqrt{x})$$

this means that $\alpha_{2k,2n}$ corresponds to the probability density $\frac{2}{\pi} \arcsin(\sqrt{x})$ (which is a catenary-like curve).

If you extract a ball (from a Bernoulli urn) every second for one year, you have more than 10% probability that before 9 days you get a parity of white and black balls and, after that, always one type (white or black) will be the winning colour during the following 356 days. Moreover, with a 5% probability last parity is reached within 54 h, and one type will always lead up to the end.

This situation shows that random symmetries are very often very particular, or not symmetric in a trivial way. In this case, the last parity follows a kind of symmetry with extremal preferences.

Randomness and Montecarlo methods

Probability is deeply linked with randomness. As we already argued in [2], the lack of information is one of the most typical features of any random process. Therefore, it is not surprising that informational analysis of purely random strings, or of probabilistic distributed strings, can benefit from informational indexes. The notion of randomness appears at the beginning of Probability Theory (Jakob Bernoulli's treatise"Ars Conjectandi sive Stochastice" can be translated as "The art of conjectures and casualty").

An important distinction has to be done, in order to understand the nature of randomness. As for probability and information, we need to distinguish between a priori and a posteriori randomness. The a priori notion of random string of length n over m symbols is very simple: it is obtained by choosing for any position of the string one of the symbols of the alphabet with probability $1/m$.

However, it is not easy to say in a rigorous and general way if a given string is random or not. This means that the *a posteriori* notion of randomness is very difficult to be defined in a general and in rigorous mathematical way. Intuitively, any string with no apparent rule in the arrangement of its symbols is random, but this is a negative definition. What appears with no rule can really follow some internal mechanism perfectly regulated, but in a way that it is almost impossible to discover, or if possible, only with large consumption of time and computational resources. Whence, a difficulty is evident in finding good definitions of *a posteriori* randomness.

In practice, the randomness of a process, and any string generated by it, is judged empirically, by statistical tests [3] trying to verify that some parameters follow probability distributions of typical natural random processes (dice, play cards, urn extractions, coin tossing, measurement errors).

We use the expression "purely random" for expressing a process of generation of values where at each step values can occur with a uniform probability distribution, whereas, in probability theory, random is used for any kind of probability distribution.

Purely Random sequences are not easy to be generated, and in nature, they can be observed in particular cases. A famous example that had an important historical role in physics is that of **Brownian motions**, observed by the botanist Brown at end of the 19th century and fully explained by Albert Einstein at beginning of the 20th century.

In physics, many purely random phenomena have been discovered. Randomness is intrinsic in the measurements of quantum systems, where physical parameters are generated by processes indistinguishable from random walks or Bernoulli processes. Therefore, randomness is a fundamental component of natural processes at the deepest level of microscopic reality.

Surprisingly, in the middle of the 20th century, another strong connection between randomness and computation was discovered: there exist mathematical functions generating, deterministically, sequences of values that have the same appearance as purely random processes of non-deterministic casualty (dice, random walks) or of physical non-determinism (Brownian motions, quantum measurements). This kind of process was called **deterministic chaos**. Both the notions of **non-deterministic casualty** and **deterministic chaos** refer to the lack of knowledge of rules generating some processes. For this reason, in practice, it is very difficult to establish if some underlying rules exist but is unknown, or if they do not exist. In both two cases, what is missing is the information linking an observed process with the causes generating it. Therefore, randomness and pseudo-randomness are hardly distinguishable.

Already at beginning of 1900, the French mathematician Henry Poincare claimed that casualty does not express the absence of causes/rules, but only our ignorance of them, or better, the impossibility of discriminating among a great number of causes where none of them is dominating over the others.

Deterministic chaos is a subtle phenomenon, because although the cause can be also very simple and due to precise deterministic rules, their effect may provide dynamics with an apparent lack of any regularity. Surprisingly, even quadratic functions provide deterministic chaos; the simplest one is the discrete map having the following recurrent form:

$$x_{n+1} = \alpha x_n(1 - x_n)$$

when α has values near 3.8 the sequence has a purely random aspect.

Linear congruential generators are defined by recurrent equations of the following form where: $[n]_m$ denotes the value of n modulus m (the remainder of the division n/m), α is the multiplier, c is the increment, x_0 is the initial value of the sequence, also called the seed of generator. Usually α and c are relatively prime positive integers:

$$x_{n+1} = [\alpha x_n + c]_m$$

$m = 2^{31}$, $\alpha = 1103515245$, $c = 12345$ are examples of values used in available generators. A theory has been developed for defining suitable values ensuring conditions for "good" pseudo-random linear congruential generators having also efficient computational performances. In these cases generated sequences pass all usual statistical tests of randomness.

A great number of algorithms are available that generate **pseudo-casual** sequences. These sequences have a huge of important applications, especially in cryptography and security protocols. Namely, the ignorance of "secret" parameters of **pseudo-casual generators** makes them a powerful source of algorithmic methods based on randomness.

The power of purely random dynamics, over a given space, is its ability to visit the space in a uniform and unbiased way. This property is mathematically expressed as **ergodicity**, and is typical of purely random, but also of deterministic chaotic processes. In simple words, ergodicity means statistical homogeneity, as a coincidence between time and probability: the development of a process in time is coherent with the probability space of the events where the process is realized. A simple application of chaos can help to give an intuition. In fact, for evaluating the size of a closed subspace in a given space, we can move along a random walk visiting the space and, after a number of steps, we count the number of points inside the subspace; the ratio m/n between the m internal points and the total number n of points is a good evaluation of the relative size of the subspace. In such evaluation the more the dynamics are chaotic, and the longer the trajectory is, the more the evaluation result is accurate.

Montecarlo methods refer to a class of algorithms that use pseudo-casual sequences for exploring spaces of events. Here we want only to explain the intuition of a method for obtaining a sequence following a given random distribution, starting from a purely random sequence. Let us consider the simple case of a finite number of values $x_1, x_2, \ldots, x_k$ that are distributed with probabilities $p_1, p_2, \ldots, p_k$ respectively. Then consider a pseudo-random generator providing values in the interval $[0, 1]$, and partition this interval into k contiguous subintervals of lengths equal to $p_1, p_2, \ldots, p_k$. At any step generate a pseudo-random number in $[0, 1]$. If the value falls in the interval j, then the corresponding value x_j is provided as output. In this way, the values $x_1, x_2, \ldots, x_k$ are surely generated according to the distribution $p_1, p_2, \ldots, p_k$. Of course, this simple idea can be extended and adapted to many possible cases, but it explains the importance of purely random processes and their centrality in the theory and applications of random variables.

Given a continuous probability density f, then the method of the **inverse cumulative distribution** can be used, for obtaining a sequence of random values generated according to f, from a purely random sequence. Namely, if F is the cumulative distribution associated with f, then F is a 1-to-1 correspondence of the interval $(0, 1)$ in itself, and the following proposition holds.

Proposition 5.5 *Given a continuous probability density f having cumulative probability distribution F, and a purely random sequence of values $(x_1, x_2, \ldots x_n)$ generated by a uniform random variable X, then the sequence:*

$$(F^{-1}(x_1), F^{-1}(x_2), \ldots F^{-1}(x_n))$$

is generated according to the probability distribution f.

Proof Let us consider the variable $F^{-1}(X)$. Given an interval $[a, b]$, being F a 1-to-1 continuous function, the probability that $F^{-1}(X) \in [a, b]$ is the same of $F(F^{-1}(X)) \in [F(a), F(b)]$, which equals the probability of $X \in [F(a), F(b)]$. But X is uniform, whence this probability is $F(b) - F(a)$. In conclusion, the probability of $F^{-1}(X) \in [a, b]$ is $F(b) - F(a)$, which implies that $F^{-1}(X)$ is distributed according to the cumulative probability density F, that is, according ton the probability density f. □

Other methods exist, which are very efficient, one of them is the Metropolis-Hastings algorithm (based on *Markov chains*, see later on) [4].

First digit phenomena and Benford's Law

In 1881, a paper appeared in the American Journal of Mathematics, the *first-digit phenomenon* was reported, according to which logarithmic tables are used with a probability that is not uniform, but seems to follow the empirical probability distribution $p_d = \log_{10}(1 + 1/d)$, where d is the first digit of a decimal logarithm and p_d is the probability of occurring this number in the context of logarithmic tables applications.

The same kind of observation was confirmed in 1938 by the physicist Frank Benford, and the appearance of this strange phenomenon was observed to happen in other contexts (even with different systems of number representations). In 1961 Roger Pinkham proved that this phenomenon is due to an interesting property of some random variables: their **scale invariance**. Let us show briefly this fact according to [5].

A random variable X is scale-invariant if the probability of a X value, within a given interval, is the same after multiplying the interval by a scale factor:

$$P(y < X < x) = P(ay < X < ax)$$

in terms of probability density $\varphi(x)$, by differentiating we obtain:

$$\varphi(x) = a\varphi(ax)$$

therefore:

$$\varphi(ax) = \frac{1}{a}\varphi(x).$$

Let us assume that X is scale-invariant and let Y be the random variable $Y = \log_b(X)$, then the following proposition holds.

Proposition 5.6 *The logarithm of a scale-invariant random variable has a constant probability density function.*

Proof Let $y = \log_b x$, then by denoting with $\varphi(x)$ and $\varphi(y) = \varphi(\log_b(x)$ the probability density functions of X and Y, we have:

$$\varphi(\log_b x) = \varphi(x) \times \frac{dx}{dy} = \varphi(x) \times \frac{db^y}{dy} = \varphi(x) \times b^y \lg b = x\varphi(x) \lg b$$

therefore, by using the scale invariance, we obtain:

$$\varphi(\log_b ax) = ax\varphi(ax) \lg b = ax\frac{1}{a}\varphi(x) \lg b = x\varphi(x) \lg b = \varphi(\log_b x)$$

but at the same time:

$$\varphi(\log_b ax) = \varphi(\log_b x + \log_b a)$$

and equating the right members of the two last equations above:

$$\varphi(\log_b x) = \varphi(\log_b x + \log_b a).$$

Since a can be any value, the equation above implies that $\varphi(\log_b x)$ is constant. □

Now we prove that Benford's law is a consequence of the scale invariance, by using the proposition above. In fact, the probability of a number $x \times 10^k$ of having the first digit equal to n, for $n \in \{1, 2, \ldots, 9\}$, is $P(n \leq x < n + 1)$, then according to the constancy of probability density, proved by the proposition above, we have:

$$P(n \leq x \leq n + 1) =$$

$$P(\log_{10} n \leq \log_{10} x \leq \log_{10}(n + 1)) =$$

$$\log_{10}(n + 1) - \log_{10} n =$$

$$\log_{10}\left(\frac{n + 1}{n}\right) =$$

$$\log_{10}\left(1 + \frac{1}{n}\right).$$

The last equation is just Benford's law.

We conclude by observing the distribution of the first digit in the powers of 2 (expressed in decimal notation), where a clear phenomenon of the predominance of 1 and 2 is evident. The powers of 2 from 0 to 9 are:

$$1, 2, 4, 8, 16, 32, 64, 128, 256, 512$$

whereas first digit 1 occurs 3 times, 2 occurs 2 times, 3, 4, 5, 6, 8 once, and 7 and 9 do not occur. It is easy to realize that a quasi-periodical behaviour is evident because after 1024 the first digits follow a similar pattern. Namely, the dominance of digit 1, followed by digit 2 is confirmed:

$$1024, 2048, 4096, 8192, 16384, 32768, 65536, 131072, 262144, 524288$$

and again, the same type of distribution holds in the next group of powers:

$$104\ldots, 209\ldots, 419\ldots, 838\ldots, 167\ldots, 335\ldots, 671\ldots, 134\ldots, 268\ldots, 536\ldots.$$

Going on, we realize that at some step 7 occurs as the first digit and then the power of the next group starts with $111\ldots$ and digit 9 appears (instead of 8) as the first digit. In this case, the dominance of the first digit 1 is related to the fact that doubling numbers starting with digits 5, 6, 7, 8, and 9 always 1 occurs as the first digit ($59 \times 2 = 118$, $69 \times 2 = 138$, $79 = 158$, $89 \times 2 = 178$, $99 \times 2 = 198$).

Markov chains

Discrete Markov chains, introduced by Andrej Markov (1856–1922) are discrete stochastic processes, that is, sequences of states generated by starting from an initial state, where at each step a state is generated according to a probability that depends exclusively on the state of the current step. If the probability depends only on the state of the current step but does not depend on the value of the step, then the chain is **homogeneous**, and if the set of possible states is finite, then the Markov chain is **finite**.

In this section, we consider only homogeneous and finite Markov chains. In this case the probability that state i is followed by state j is given by the transition probability $p_{i,j}$ and the stochastic process of a Markov chain is completely defined by an initial probability distribution $\{p_i\}_{i \leq n}$ giving the probability that the initial state is one of the n possible states and by a stochastic square matrix of non-negative real numbers $\{p_{i,j}\}_{i,j \leq n}$, called **transition matrix**, giving the transition probabilities for each pair (i, j) of states.

Another way of describing a transition matrix is by means of a graph where nodes are states and edges connecting nodes are the probabilities of transition.

Markov chains have wide applicability in many fields. In several cases, they revolutionized the methods of solving problems. In terms of Markov chains, many important dynamical and probabilistic concepts can be formalized and analyzed. We refer to [1] for a good introduction to the general theory of Markov chains. One of the main properties of some Markov chains is their **ergodicity**, that is, the property of reaching a **limiting probability distribution**, independently from the initial distribution of the chain. In this section, we prove this property, for a class of Markov chains, as a consequence of the Perron-Frobenius Theorem. This Theorem is a result of linear algebra, obtained at beginning of 1900, with a lot of important implications in several fields. Here we consider its relevance for Markov processes.

The following is its usual formulation.

Proposition 5.7 (Perron-Frobenius Theorem) *Let A an $n \times n$ matrix of real values that are positive, that is, $A = (a_{i,j} > 0)_{1 \leq i,j \leq n}$, writing shortly $A > 0$. The following conditions hold:*

1. *$\exists \lambda_0 > 0 \in \mathbb{R}$ that is eigenvalue of A, with u_0 the corresponding eigenvector: $\lambda_0 u_0 = A u_0 \quad u_0 > 0$;*
2. *every eigenvector of A is multiple of u_0;*
3. *every eigenvalue of A, possibly in the field of complex numbers, has an absolute value inferior to u_0.*

Proof We will prove only the first of the above conditions, which is relevant for our further discussion. Let $\lambda_0 = max\{\lambda > 0 \in \mathbb{R} \mid \lambda u \leq Au, u \in \mathbb{R}^n\}$. Surely this maximum value exists because the set is closed and bounded ($Au \leq Mu$ for each vector u when M is the $n \times n$ matrix such that $max(a_{i,j})_{1 \leq i,j \geq n}$ is put in all positions). Of course $\lambda_0 u_0 \leq A u_0$. We will show that $\lambda_0 u_0 = A u_0$. In fact, let us assume that equality does not hold: $\lambda_0 u_0 < A u_0$, then by the assumption $A > 0$ we obtain:

$$A(Au_0 - \lambda_0 u_0) > 0$$

whence by setting $y_0 = Au_0$:

$$Ay_0 > \lambda_0 u_0.$$

Therefore for some $\lambda' > \lambda_0$:

$$Ay_0 > \lambda' u_0$$

that is an absurd because by definition λ_0 is the maximum value satisfying the inequality. In conclusion $Ay_0 = \lambda_0 u_0$. Moreover, $A > 0$ implies that when $Au_0 = \lambda_0 u_0 > 0$, being $\lambda_0 > 0$, necessarily $u_0 > 0$. □

A matrix of real values is *left stochastic* when the sum of its values over any column is equal to one.

Corollary 5.1 *If B is a left stochastic real matrix, then* $\lambda_0 = 1$.

Proof Let $B = (b_{1,j})_{1 \le i,j \le n}$. We have that for an eigenvalue λ_0:
$\sum_j b_{j,1} u_j = \lambda_0 u_1$
$\sum_j b_{j,2} u_j = \lambda_0 u_2$
....
$\sum_j b_{j,n} u_j = \lambda_0 u_n$
therefore, by summing, for $1 \le j \le n$:
$\sum_j (b_{j,1} + b_{j,2} + \ldots + b_{j,n}) u_j = \lambda_0 u_j$
but, from the left stochasticity of B it follows $(b_{j,1} + b_{j,2} + \ldots + b_{j,n}) = 1$ therefore $\lambda_0 = 1$. □

A real matrix A is primitive if for some m the power matrix A^m has all components > 0. Therefore, for a primitive stochastic matrix B, for all $k > m$ matrices B^k have all components < 1, because if some value of the B were equal to 1, then all the remaining values of the column have to be null, but this contradicts the fact that in B^n all the components are strictly positive.

Corollary 5.2 *Given two stochastic vectors* $u, v \in \mathbb{R}^n$ *(with components giving sum* 1*) and a primitive stochastic matrix A, then for some* $m \ge 1$ *and for any* $k > m$ $|A^k(u - v)| < 1$.

Proof If for both u, v, $u[i] < 1$ and $v[i] < 1$ for $1 \le i \le n$, then being also $u > 0$ and $v > 0$, the asserted inequality surely holds because A is stochastic and primitive. The same situation holds even if one component of u or v equals 1 (at most one component can have this value in each vector). In the case that for two values $i \ne j$ both equations $u[i] = 1$ and $v[j] = 1$ hold, then $(u - v)[k] = 0$ for $k \ne i, k \ne j$ whereas $(u - v)[i] = 1$ and $(u - v)[j] = -1$ or *vice versa*. In this case, when A^k is multiplied by $(u - v)$, we get as result two columns of A^k, that are stochastic vectors with all components less than 1, therefore we are in the same situation as the first case. In conclusion, in all cases the inequality $|A^k(u - v)| < 1$ holds. □

Now let u_0 be the eigenvector postulated by the Perron-Frobenius Theorem of a primitive stochastic matrix B^m. According to Corollary 5.1, the eigenvalue of u_0 is 1, whence $Bu_0 = u_0$. Thus, let us consider any stochastic vector v different from u_0, we claim that:

$$\lim_{m\to\infty} B^m(v - u_0) \to 0.$$

In fact,

$$B^m v = B^m(u_0 + (v - u_0)) = B^m u_0 + B^m(v - u_0)$$

but $Bu_0 = u_0$ implies that $B^m u_0 = u_0$, while for $m \to \infty$ the value $B^n(v - u_0)$ approaches to zero, as m increases, because all the components of B^m are < 1 and, from Corollary 5.2, for some k and any $m > k$ also $|B^n(v - u_0)| < 1$.

The discussion above has a deep consequence in the theory of Markov chains. In fact, the conclusion we obtained above, as a consequence of the Perron-Frobenius Theorem, tells us that starting from any stochastic vector, if the stochastic matrix is primitive, we reach a limit state probability that does not depend on the initial state probability and corresponds to the eigenvector of the matrix. This possibility for Markov chains is the basis of an important algorithm due to Metropolis-Hastings [4] and based on pseudo-random numbers, the MonteCarloMarkovChain algorithm, by means of which is possible to evaluate probabilities even when the analytical form of density distribution is difficult to express or to integrate analytically.

Google Page Rank Algorithm

Here we present an elegant application of the Perron-Frobenius Theorem to the Google Page Rank Algorithm [6]. Let us consider a Google query. The way the links of relevant pages are listed is a crucial aspect of Google's research engine. Firstly, all the web pages containing the words of the query are determined, and then these pages are organized as a graph where pages are the nodes and two pages are connected with an oriented edge if the first page contains an HTML link to the second page. Let us include the initial page P_0 as a page having an edge toward any of the selected pages. Let $B = (b_{i,j})$ the boolean matrix associated with the graph we constructed, and let us denote by $O(i)$ the number of edges exiting from the edge i.

Now from the matrix of the graph, of nodes $P_0, P_1, P_2, \ldots, P_n$, we define a stochastic matrix C, by fixing a threshold value $p < 1$, usually set as 0.75, which we call *search probability*. The components of matrix $C = (c_{i,j})$ are defined in the following way.

$c_{0,i} = 1/n$
$c_{i,0} = 1$ if $O(i) = 1$
$c_{i,0} = 1 - p$ if $O(i) \neq 1$

This means that from the initial page, we can move with the same probability to any of the n pages, whereas if P_i is not connected with no page different from P_0, then we

necessarily go to P_0, and if P_i is connected with $O(i)$ pages, we have a probability $1 - p$ of starting again the search from P_0.
Moreover, from each page we can go to any connected page with the same probability, the search probability divided by the number of connected pages:

$c_{i,j} = 0$ if $b_{i,j} = 0$
$c_{i,j} = p/O(i)$ if $b_{i,j} = 1$.

The matrix C is stochastic and it is also primitive because two nodes are always connected by a path, that is, the probability of going from one node to any other node, in some number of steps, is different from zero (possibly passing through P_0). This means that there exists m such that $C^m > 0$. Therefore, we can deduce, according to Perron-Frobenius Theorem, that if u_0 is the eigenvector of C, then starting from any stochastic vector v, by applying iteratively C we approach to the limit vector u_0. This vector approximates the probability of reaching a given page along a very long path in the graph. For this reason, the Brin-Page algorithm of Page-Rank lists the pages connected to P_0 according to the decreasing order of the probabilities assigned by u_0.

Topologies, Measures, Distributions

The general mathematical setting of probability is based on the mathematical theory of measure that was developed since the seminal research of Lebesgue, Borel, and other mathematicians, especially of the French school, at beginning of the twentieth century. In general, the notion of measure is formulated within a topological setting, and probability is a special case of measure.

A **topological space** over a set S is given by associating to any element x of S a family $U(x)$ of non-empty subsets of S that are called **neighborhoods** of x. These neighbourhoods express the intuition of localization of x in S by means of an approaching "telescopic" sequence of sets including x in an increasingly closer way. In fact, it is assumed that $S \in U(x)$ for every $x \in S$, and if $V \in U(x)$ then $x \in V$ and exists $V' \in U(x)$ such that $V \supset V'$; if both $V_1 \in U(x)$ and $V_2 \in U(x)$ then $V_1 \cap V_2 \in U(x)$; and finally, if $V \in U(x)$ there is $W \in U(x)$ such that $V \in U(y)$ for every $y \in W$. An **open set** of a topology is any set that is a neighbourhood of all its elements, while a **closed set** is complementary of an open set.

The notion of limit and continuous function can be defined in general topological terms without assuming any notion of distance. A function f between two topologies is continuous when the f-inverse image of any open set is an open set too. For this reason, a topology allows us to define the classical notions of mathematical analysis in more general and abstract terms, by individuating the essence of many crucial spatial concepts. We remark that the definition of continuity by means of inverse images is stronger than a possible "direct" definition. In fact, in inverse terms, continuity requires that points that are close together cannot be images of points that are not close together, that is not only "vicinity" is conserved by a continuous function, but "non-vicinity" in the domain cannot be transformed into vicinity in the codomain.

A collection of subsets of a set S that is closed under taking complements and finite unions is called a **field** (or an algebra of sets). A field that is closed under taking countable unions is called a σ-field. A Borel σ-field is a σ-field generated by the open sets of a topology (the smallest σ-field containing all the open sets).

A **measurable space** is given by (S, A) where A is a σ-field of subsets of the set S including $\emptyset$.

Given a measurable space (S, A), then a **measure** μ on S is a function from A to the non-negative real numbers extended (with the infinite value ∞) that satisfies:

(1) $\mu(\emptyset) = 0$,
(2) the countable union of mutually disjoint measurable sets is the (countable) sum of the measures of the sets.

The triple (S, A, μ) is called a **measure** space. If (S, A, μ) is a measure space and $\mu(S) = 1$, then μ is called a probability and (S, A, μ) is called a **probability space**.

Suppose that (S, A) is a measurable space, and let (T, C) be another measurable space. A **measurable function** $f : S \rightarrow T$ is a function such that for every subset B that belongs to C the inverse image $f^{-1}(B)$ belongs to A.

Given a probability space (S, A, μ), any element of A is called an **event**. A measurable function X from S to the set C of some other measurable space (C, B) is called a random quantity. A **random quantity** X is a **random variable** when C is the set $\mathbb{R}$ of reals and B the Borel σ-field over $\mathbb{R}$.

The probability measure μ_X induced on (C, B) by the random quantity X from μ is called the **distribution** of X. In other words, a distribution is a measure determined by means of inverse images from a previously assumed measure. However, apart from this very general setting, the intuition behind the concept of distribution is that of a quantity that is distributed among the individuals of a population. In this sense, it is a truly statistical concept.

5.2 Informational Randomness

In a previous section, we defined random strings, as those generated by Bernoulli processes with the uniform probability of extracting the different values of a finite variable. This is a good definition for generating random strings. But when a string is given how can we establish if it is random, that is, if it was generated by a sort of Bernoulli process or not?

In this section we will give a very simple rule for answering to the question above. Given a string α of length n over m symbols, let us consider the shortest length ∇ such that all the strings of length ∇ that occur in α occur only once in it. If

$$\nabla = 2 \log_m n$$

or it is very close to this value, then α is almost surely a string randomly generated over its m symbols.

The results that we present in this section were developed in relation to genomes and random genomes, which are very important for discriminating meaningful biological aspects from those that have a purely casual nature. However, their relevance is very general, therefore we present them in a chapter devoted to general concepts of information and probability. Often generic strings will be indicated by $\mathbb{G}$, even when they are strings over alphabets that are not genome alphabets. The results given in this section improve those presented in [2, 7, 8].

A deep link between information and randomness was discovered within *Algorithmic Information Theory* [9, 10] developed, by Andrej Nikolaevič Kolmogorov (1902–1987) and Ray Solomonoff (1926–2009), and by Gregory Chaitin (1947–), in a more computer science oriented setting. According to this theory the **algorithmic information** of a string is given by the length of the shortest program able to generate the string (in a fixed programming language). Then, strings having **algorithmic information** greater than or equal to their lengths are **algorithmically incompressible** and for this, they are considered **casual** or **random** (no efficient deterministic rule exists for generating them).

This perspective is illuminating in many aspects. It sheds new light on the connection between *comprehension* and *compression*, which have the same Latin root, because the *comprehension* of a phenomenon is a form of reduction of some data observed about the phenomenon to a few principles of a theory within which the phenomenon is interpreted. Moreover, in the algorithmic perspective, a clear link is established between information, randomness, and computation.

From a practical viewpoint, the algorithmic notion of casualty cannot be recognized in a complete exact way, because the property of a string of being algorithmically casual is not computable. Thus, you can define the property, but you cannot show, in formal terms, that the property holds for a given string.

This limitative result is of a very general nature. In fact, other mathematical formalizations of randomness for finite, or infinite, strings, outside the probability theory, suffer a sort of incompatibility between a rigorous definition of randomness, based on intrinsic properties of strings, and the possibility of verifying these properties in a computable way.

In a paper by Martin Löf [11] a sequence is casual when it is **generic**, that is, when all Turing machine tests for checking specific properties fail. This implies that casualty is not a decidable property. In fact, assuming its decidability, then a Turing machine would be able to decide if a given sequence is casual, therefore the sequence does not fail for a particular computable test, whence the sequence could not be casual, thus reaching a contradiction.

The only way to approach a characterization of randomness can be obtained only in probabilistic terms or within some degree of approximation. The probabilistic approach is already well-established and it is the basis for the actual statistical tests, organized in suitable packages, used in the majority of applications. In the sequel, we will explain our informational, approximate criterion of randomness for strings.

A randomness index based on hapaxes

Let us indicate by $RND_{m,n}$ random strings, of length n, over an alphabet of m symbols, that is, a sequence of length n generated by a random process where at each step one of m symbol is generated with a probability of $1/m$.

The strings in $RND_{m,n}$ obey to an **Homogeneity Principle** very similar to *normality* introduced by Emil Borel for positional representations of irrational numbers (infinite sequence of digits) [12, 13]. According to Borel, an irrational number is said to be normal if, for every positive integer n, all strings of n digits have the same probability of occurring in the representation of the number in a given base. It can be shown that any irrational number is normal, and normality is assumed as a form of randomness. Analogously, we assume that strings of $RND_{m,n}$ obey to an **internal homogeneity**, according to which, in any random string, all portions of a given length have the same probabilistic properties (properties expressed by means of conditions on probabilities). Moreover, strings of $RND_{m,n}$ obey also to an **external homogeneity**, according to which all random strings, singularly considered, satisfy the same probabilistic properties.

We recall that $\nabla(\mathbb{G})$ is the no-repeat of a the string $\mathbb{G}$. The following proposition tells us that if $k = \nabla(\mathbb{G})$, then the E_k-entropy of $\mathbb{G}$ is the maximum among the E_h-entropies of $\mathbb{G}$ for values grater than k.

Proposition 5.8 *Given a string $\mathbb{G}$ of length n over m symbols, if $k = \nabla(\mathbb{G})$, then $2\lg_m(n-k+1)$ is the maximum value that $E_k(\mathbb{G})$ reaches in the class of values $\{E_h(\mathbb{G}) | n \geq h \geq k\}$.*

Proof Minimum value of k such that all k-mers are hapaxes of $\mathbb{G}$ is $\nabla(\mathbb{G}) = mrl(\mathbb{G}) + 1$ (mrl is the maximum length of repeats). Hence, for $k = \nabla(\mathbb{G})$ empirical entropy $E_k(\mathbb{G})$ is equal to $\log_m(n-k+1)$, in fact $(n-k+1)$ is the number of distinct k-mers in $\mathbb{G}$. The same expression holds for any $h \leq n$ and $h \geq k$, because strings longer than k are hapaxes too. But if $h > k$ then $\log_m(n-k+1) > \log_m(n-h+1)$, therefore $E_k(\mathbb{G})$ is the maximum of $E_h(\mathbb{G})$ in the interval $n \geq h \geq k$. □

In the first half of 1900, Nicholas de Brujin (1918–2012) introduced a class of graphs generating, for any $k > 0$, **cyclic sequences** of length $m^k - k - 1$ over m symbols that are k-**hapax**, called de Brujin sequences of order k, where all possible k-mers occur. Such sequences are cyclic because they are arranged on a circle, where the k-mer ω at the end of a de Bruijn sequence is followed by $k-1$ other k-mers that begin at a position i of ω, with $1 < i \leq k$, and continue at positions $1, 2, \ldots, i-1$ of the sequence.

A de Bruijn graph has all the possible strings of length k as nodes, where an oriented edge from node α to node β is present if the $(k-1)$-suffix of α is equal to the $(k-1)$-prefix of β. Therefore, each node has m exiting edges and m entering edges. In this way, starting from a node, any path of $n-1$ edges connects $n-k+1$ nodes that, when concatenated with $(k-1)$ overlap, provide a string of length n.

All the nodes of a de Bruijn graph have even degree ($2m$ if strings are over m symbols). For this reason, it can be shown that there exists a path passing for all the nodes once only once (Hamiltonian path), and also passing once only once through all edges (Eulerian path). The existence of this path ensures the existence of a de Bruijn sequence where all possible k-mers occur and where each of them occurs only once in the sequence.

The following proposition generalizes a result obtained for random genomes [7, 8]. It gives an evaluation of the no-repeat bound in random strings.

Let us recall that $\nabla_{m,n}$ denotes the no-repeat length for any string of length n over m symbols. Namely, according to the homogeneity of random strings, this value has to be the same for all the strings of $RND_{m,n}$.

Proposition 5.9 *Let* $\nabla_{m,n}$ *be the no-repeat length for strings of* $RND_{m,n}$*. The following* **logarithm randomness condition** *holds:*

$$\nabla_{m,n} = \lfloor 2\log_m(n) \rfloor$$

Proof There is in R a string of length $\nabla_{m,n}$ which is an hapax of R. Therefore, from the homogeneity of random strings, all $\nabla_{m,n}$-mers of R are hapaxes of R. In other words, any $R \in RAND_{m,n}$ is a $\nabla_{m,n}$-hapax string.

According to de Bruijn graph theory, we know that we can arrange, for any $k > 0$, all m^k possible k-mers in a circular sequence α (the last symbol of α is followed by the first symbol of α) where each k-mer occurs exactly once. Of course, any contiguous portion long n of α contains $(n - k + 1)$ consecutive k-mers and corresponds to a string, which we call a n-portion of α, where all substrings of α with lengths less than or equal to k are equi-probable, and this probabilistic homogeneity holds along all positions of α, in the sense that, going forward (circularly) other be Brujin sequences, with the same equi-probability property are obtained.

Now, let α be a de Bruijn sequence arranging all k-mers where $k = \nabla_{m,n}$. Let us consider the disjoint n-portions (with no common k-mer) that are concatenated in α. Their number r has to be equal to $m^k/(n - k + 1)$, and any n-portion belongs to $RND_{m,n}$, given its homogeneous probabilistic structure. But α is a circular string arranging all k-mers, where any k-mer occurs once in α, therefore, any k-mer has to belong to some n-portion of α and only to one of them. Moreover, each k-mer which belongs to a n-portion occurs only in one of its $(n - k + 1)$ positions. Therefore, the maximum probability homogeneity is realized when $r = (n - k + 1)$, that is, when the number r of n-portions of α is equal to the number of positions of any n-portion. In fact, this implies that any k-mer of m symbols occurs in some random n-portion of α with the same probability of occurring in any position of such random portions.

In conclusion, if $k = \nabla_{m,n}$, then:

$$m^k/(n - k + 1) = (n - k + 1)$$

that is:

$$m^k = (n - k + 1)^2$$

giving the following equations and inequalities:

$$k = 2\log_m(n - k + 1)$$

$$k = 2\log_m(n - 2\log_m(n - k + 1) + 1)$$

$$2\log_m(n - 2\log_m(n)) < k < 2\log_m(n)$$

$$2\log_m(n - \frac{m-1}{m}n) < k \leq \lfloor 2\log_m(n) \rfloor$$

$$2\log_m(n/m) < k \leq \lfloor 2\log_m(n) \rfloor$$

$$2(\log_m(n) - \log_m(m)) < k \leq \lfloor 2\log_m(n) \rfloor$$

$$2\log_m(n - 1) < k \leq \lfloor 2\log_m(n) \rfloor$$

$$\lfloor 2\log_m(n) \rfloor \leq k \leq \lfloor 2\log_m(n) \rfloor$$

thus, the last inequality implies the equation asserted by the proposition. □

The proposition above suggests the following hypothesis.

Hypothesis of informational randomness The class $RND_{n,d}$ coincides with the class of strings of length n over m symbols where $\nabla_{m,n} = \lfloor 2\,\mathrm{lg}_m(n) \rfloor$.

Let β a string of length n over m symbols that differs only in one position from a string $\alpha RAND_{m,n}$. Is β a random string or not? And, what about its randomness, if it differs from α in two, or three, or more positions?

Intuitively, it seems reasonable that for a string of millions or billions of occurrences, very small differences do not alter the random character of the whole string. This kind of situation arises in many contexts. For example, what is a "pile of stones"? If from a pile of stones we remove a stone, does remain it a pile of stones? These questions cannot have precise answers, because are based on intrinsically imprecise concepts. Therefore, the only way to answer adequately to them is according to some approximation degree. In this sense, there is an agreement with the probabilistic notion of randomness, defined in terms of the statistical tests passed by a given string, and on the corresponding degrees of fulfilment of these tests.

In conclusion, randomness is not a 0–1 property, but what is also called, after Lotfi Aliasker Zadeh (1921–2017), a *fuzzy* property, in the sense of fuzzy logic, which holds with a value between 0 and 1. Therefore, a string is informational random in a degree that depends on the closeness of its ∇-index to the value $\lfloor \log_m(n) \rfloor$. Of course, several criteria for measuring such a degree can be given, suitable to specific contexts of application.

Table 5.1 shows the results of an experiment based on Proposition 5.9, confirming $2\,\mathrm{lg}_4(n)$ as a good approximation of the average value of $\nabla_{n,4}$ in $RND_{n,4}$.

Results similar to those given in the table above were obtained with random sequences generated by quantum phenomena, games, irrational numbers, pseudo-random generators [3, 14–16], by obtaining an agreement, within a very good level of approximation, between randomness, pseudo-randomness and informational ran-

Table 5.1 Pseudo-random strings over 4 symbols of different lengths n have been generated (100 strings for each length); their $\nabla_{4,n}$ values have been evaluated, by reporting in the table, for each length, the minimum $\nabla_{n,4}$ value (min), the maximum $\nabla_{4,n}$ value (max), standard deviation (sd), and the average $\nabla_{4,n}$ value (avg). With a good approximation it results that $2\lg_4(n) = \log_2 n \approx avg$

Length	Min.	Max.	Sd.	Avg.	$lg_2(n)$
1,000	9	15	1.07	10.2	9.97
100,000	15	20	0.95	16.67	16.61
200,000	16	21	0.86	17.78	17.61
500,000	18	23	0.91	19.09	18.93
1,000,000	18	24	0.96	20.14	19.93
10,000,000	22	26	0.97	23.49	23.25
20,000,000	23	27	0.93	24.31	24.25
30,000,000	24	30	1.14	25.08	24.84
50,000,000	24	31	1.17	25.86	25.58
75,000,000	25	29	0.85	26.44	26.16
100,000,000	25	30	1.02	26.89	26.58

domness [2]. In conclusion, the definition of informationally random string, given above, as $2\log_m(n)$-hapax string, seems to be quite satisfactory, and very useful in many contexts.

In the perspective of Proposition 5.9, de Bruijn graphs can be considered a class of generators of pseudo-random strings of given lengths.

The following proposition shows that $\lfloor 2\log_m(n)\rfloor$ corresponds to an upper bound to the entropy of strings of length n over m symbols, and $2\log_m n$ is essentially a fix-point for the function that associates, to any value k, the maximum value reached by k-entropy.

Proposition 5.10 *In the class of strings $\mathbb{G}$ of length n over m symbols, for any k such that $\nabla(\mathbb{G}) \leq k < n$ the following equation holds:*

$$E_k(\mathbb{G}) \leq \lfloor 2\lg_m(n)\rfloor. \tag{5.13}$$

moreover, random strings $RAND_{m,n}$ have $\nabla_{m,n}$-entropies that differs from the upper bound: $\lfloor 2\lg_m(n)\rfloor$ less than di $2\lg_m(n/(n - \lfloor 2\lg_m(n)\rfloor))$, which approximates to zero for increasing values of n.

Proof From Proposition 5.8, for $\nabla(\mathbb{G}) \leq k < n$, entropy $E_k(\mathbb{G})$ reaches its maximum value when $k = \nabla(\mathbb{G}) = mrl(\mathbb{G}) + 1$. In such a case:

$$E_k(\mathbb{G}) = 2\lg_m(n - k + 1) = 2\lg_m(n - mrl(\mathbb{G})) \tag{5.14}$$

hence, difference $2\lg_m(n) - E_k(\mathbb{G})$ is given by:

$$2\lg_m(n) - 2\lg_m(n - mrl(\mathbb{G})) = 2\lg_m(n/(n - mrl(\mathbb{G}))). \quad (5.15)$$

According to Proposition 5.9, when $\mathbb{G} \in RAND_{m,n}$ $mrl(\mathbb{G}) = \lfloor 2\lg_m(n) \rfloor - 1$, whence, by replacing in Eq. (5.15) $mrl(\mathbb{G})$ by $\lfloor 2\lg_m(n) \rfloor - 1$ we get:

$$2\lg_2(n/(n - mrl(\mathbb{G}))) < 2\lg_2(n/(n - \lfloor 2\lg_m(n) \rfloor)).$$ □

The following proposition gives an evaluation of the (enormous) number of random strings, by using the number of de Bruijn sequences.

Proposition 5.11

$$|RND_{m,m^k}| = (m!)^{(m^{2k}-1)}$$

Proof From de Brujin theory we know that the number of de Bruijn sequences of order k over m symbols is given by the following formula:

$$\frac{(m!)^{m^{(k-1)}}}{m^k}$$

whence, from the hypothesis of informational randomness, the number of random strings of length n over m symbols is easily obtained by replacing k by $2k$ and considering that, at any position of a de Bruijn sequence of order $2k$, a contiguous portion of m^k positions starts, which is a random string. Whence, by easy calculation the asserted formula follows. □

Randomness is crucial for life, as we have already seen in the chapter on genomes. It is also crucial for Physics. Quantum physics can be properly developed only within a framework where random processes are intrinsic to particle behaviours. Measurements are operators that when applied to quantum systems get values, that is, measures of quantum variables, by inducing systems to reach some specific states. However, without such interventions systems assume states that are random combinations of a huge of possible states. In this perspective, uncertainty results to be a way of parallel elaboration for a whole class of possible solutions, among which the right ones emerge. The computational power of quantum computing resides essentially in this possibility. Hence, randomness is a new frontier of computing [17–19]. Random algorithms, already discovered in computer science, are formidable instruments of problem solving, and pseudo-random numbers are ubiquitous [4, 14, 15] in many application fields of Computer Science. In recent approaches of machine learning and deep neural nets, the combination of calculus and pseudo-random processes discloses new perspectives for Artificial Intelligence, where the essence of intelligence appears just in the way randomness is employed within strategies able to elaborate goals and resolution criteria. These perspectives open new scenarios for many disciplines, by overcoming intrinsic epistemological limitations of the present scientific conceptual apparatus.

5.3 Information in Physics

In this section, three fundamental passages will be considered, which were crucial in the development of physics, and where information results to be, implicitly or explicitly, the essence for novel comprehensions of physical phenomena [20–25]. An "informational" line links Boltzmann's entropy with Planck's discovery, which extends Boltzmann's approach, and with Schrödinger's equation, which uses Planck's relation between frequency and energy and de Broglie's wave-particle duality.

Boltzmann's Equation

In 1824 Sadi Carnot wrote a book about the power of fire in producing energy [26]. It was known that heat can provide mechanical energy. Moreover, as it was experimentally later shown by Joule, mechanical energy can be entirely transformed into heat (first law of thermodynamics). However, it is impossible entirely transform heat into mechanical energy. In fact, this is a way to formulate the second law of thermodynamics, which is also equivalent to the impossibility of spontaneous processes where heat passes from a colder body to a hotter body. Carnot introduced some ideal heat engines, called reversible engines, reaching the optimal efficiency in heat-work transformation. In this way, he proved a theorem giving an evaluation of the heat quantity that necessarily cannot be transformed into mechanical energy. Namely, when an engine M takes a quantity Q_2 of heat from a heat source (boiler) at constant temperature T_2, then a quantity Q_1 has to be released to a colder body (condenser) at temperature T_1 ($T_2 > T_1$), and only the difference $Q_2 - Q_1$ can be transformed into mechanical work. For reversible engines working between temperatures $T_2 > T_1$, the released heat quantity Q_1 reaches the minimum (positive) value, and the following equation holds:

$$\frac{Q_1}{T_1} = \frac{Q_2}{T_2} \tag{5.16}$$

therefore, if we denote by S the heat quantity Q_1 when T_1 is the unitary temperature and $T_2 = T$ we obtain:

$$S = Q/T. \tag{5.17}$$

S corresponds to a thermodynamical quantity, later called by Clausius [20] the entropy (of M), corresponding to the minimum heat quantity that necessarily a heat engine, working between temperatures T and 1 has to release (to the unitary condenser) and that cannot be transformed into mechanical work.

The famous formulation of the second law of thermodynamics, by means of entropy S, asserts that in a closed system (that does not exchange energy with the external environment) the entropy cannot decrease in time.

In this scenario, in the 1870s years, Ludwig Boltzmann started research aimed at explaining the second law of thermodynamics in terms of Newtonian mechanics [21]. The main question was: "Where does the time arrow come from?". In fact, in mechanics, all the laws are symmetric with respect to time and the same equations tell us what happens in the future, but also what happened in the past. In no equation, there is an explicit indication of the direction of events in time (the first Chapter of Wiener "Cybernetics" [24] is devoted to the different notions of time in Newtonian mechanics and in biology).

The first step of Boltzmann's project was a mechanical formulation of entropy (the argument is presented according to [27]). This formulation starts from the fundamental law of ideal gases, where P is the pressure, V the volume, T the absolute (Kelvin) temperature, N the number of gas moles, and R is the gas constant:

$$PV = NRT. \tag{5.18}$$

If we pass from the gas moles N to the number of molecules n in the gas (by the relation $Na = n$ where a is the Avogadro constant), we get an equivalent formulation, where $k = Ra$ is now called the Boltzmann constant:

$$PV = nkT. \tag{5.19}$$

Now, let us assume that the gas takes some heat by expanding from a volume V_1 to a volume V_2. Then, the quantity Q of this heat is given by:

$$Q = \int_{V_1}^{V_2} P dv \tag{5.20}$$

and by expressing P according to Eq. (5.18), we get:

$$Q = \int_{V_1}^{V_2} nkT/V dv = nkT \int_{V_1}^{V_2} 1/V dv = nkT(\ln V_2 - \ln V_1). \tag{5.21}$$

Let's assume to start from a unitary volume $V_0 = 1$. If in Eq. (5.21) $V_1 = V_0$, $V = V_2$ and T are moved to the left member, then we obtain:

$$Q/T = nk \ln V \tag{5.22}$$

that, according to Carnot's equation (5.17), gives:

$$S = nk \ln V \tag{5.23}$$

that is:

$$S = k \ln V^n \tag{5.24}$$

where V^n expresses the number of possible ways of allocating the n molecules in V volume cells. The passage from constant R to constant k and from N moles to n molecules, accordingly, is crucial to the microscopic reading of the formula (very often this is not adequately stressed when Boltzmann's argument is analyzed).

We can assume that the gas is spatially homogeneous, that is, the volume cell positions do not matter and molecules are indiscernible when they have the same velocities. Moreover, only m velocity intervals can be distinguished, at a discreteness level Δ, that is, intervals $v_1 \pm \Delta, v_2 \pm \Delta, \ldots, v_m \pm \Delta$. Hence, a distribution of n molecules in the volume V is given, apart from multiplicative constants, by the number W of different ways in which n molecules can be distributed into m different velocity classes where $n_1, n_2, \ldots n_m$ are the number of molecules in each velocity classes (W is the number of different micro-states associated to a given thermodynamic macro-state). Thus S is given by:

$$S = k \ln W \tag{5.25}$$

hence:

$$\lg W = k \ln \frac{n!}{n_1! n_2! \ldots n_m!} \tag{5.26}$$

whence, by using Stirling's approximation:

$$\lg W = kn \ln n - k[n_1 \ln n_1 + n_2 \ln n_2 + \cdots + n_m \ln n_m] \tag{5.27}$$

that, apart from an additive constant and the sign, is the H function for which Boltzmann tried to prove the H Theorem asserting that is not increasing in time (in isolated systems).

Now if in the above equation we replace $n_1, n_2, \ldots, n_m$ by $np_1, np_2, \ldots, np_m$ ($p_1, p_2, \ldots, p_m$ the frequencies n_i/n) we obtain:

$$\lg W = kn \ln n - k[np_1 \ln np_1 + np_2 \ln np_2 + \cdots + np_m \ln np_m] \tag{5.28}$$

whence:

$$\lg W = kn \ln n - kn[p_1(\ln n + \ln p_1) + p_2(\ln n + \ln p_2) + \cdots + p_m(\ln n + \ln p_m)] \tag{5.29}$$

that is:

$$\lg W = kn \ln n - kn \ln n + -kn[p_1 \ln p_1 + p_2 \ln p_2 + \cdots + p_m \ln p_m] \quad (5.30)$$

or:

$$\lg W = -kn[p_1 \ln p_1 + p_2 \ln p_2 + \cdots + p_m \ln p_m] \quad (5.31)$$

showing the equivalence (apart from multiplicative constants) between the physical entropy S and the informational entropy H_S (subscript S stands for Shannon initial):

$$\lg W = -knH_S. \quad (5.32)$$

Equation (5.25) is reported in Boltzmann's tomb in Wien. In this form, the equation was later given by Max Planck, who followed Boltzmann's approach in his famous conference on December 14, 1900, from which Quantum Theory emerged [22]. The importance of this microscopic formulation of Entropy is the statistical approach that stemmed from it and that became crucial in the whole physics of the twentieth century. The computation of W was obtained from Boltzmann by means of the so-called Maxwell-Boltzmann statistics, that is the multinomial representation of W, expressed by Boltzmann as the "Wahrscheinlichkeit" principle.

From the statistical formulation of entropy, Boltzmann tried to prove the statistical nature of time arrow in thermodynamics. The question he posed was: Why does entropy increase (or better does not decrease) in time? We know that molecules colliding in a gas follow Newtonian mechanics, where time does not show a preferential direction because collisions are ruled by equations. Therefore time irreversibility has to emerge from a population reversible elementary processes. How can happen such a kind of phenomenon? Boltzmann proved in 1972 his H Theorem, showing that the function H, in an isolated system (which does not exchange energy with its environment) does not increase in time (entropy which is essentially given by -H does not decrease). In this way, the second principle of thermodynamics is a consequence of the microscopic representation of entropy. However, Boltzmann's proof was not accepted by many scientists of that time (some technical inadequacies were present in the proof). Nevertheless, Boltzmann was right, and today many proofs are available of H Theorem. Here we present a numerical experiment, which, in a very simple manner, explains why time direction is a consequence of the laws of large numbers in random processes and of Shannon entropy properties [27, 28].

Let us consider a sort of two-dimensional gas, that is, a population of numbers randomly chosen in a numerical interval (any number can occur in many copies). Let us assume specific approximation levels, by considering equal numerical values differing within a fixed error interval. Then start a process of collisions, where these numbers, seen as velocities of particles in two dimensions, evolve, step by step, by exchanging in each collision their x, y-components (according to a randomly chosen collision direction). In this way, both velocity distributions of x-component and y-

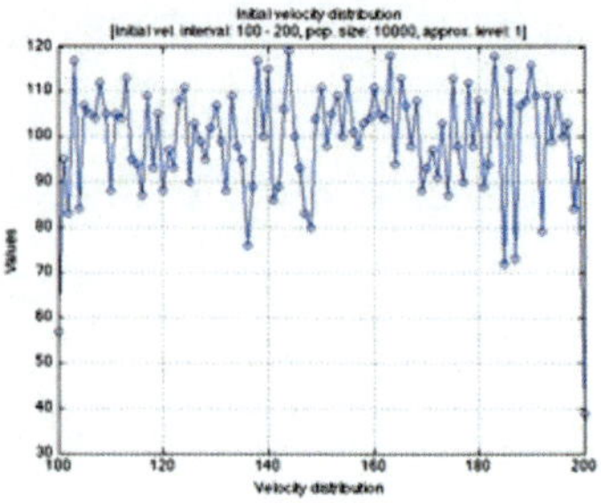

Fig. 5.1 Initial distribution, Interval 100–200, population size 10000 (from [28])

component tend, in time, to normal distributions. From Information theory, we know that a normal distribution is one having the maximum value of Shannon entropy among all distributions with a fixed variance. But in the considered case, collisions exchanging x, y-components (along a collision direction) do not change the variance of x, y-distributions (collisions are elastic, without loss of kinetic energy). Therefore, tending this distribution to normal laws, and being the numbers of the form $\sqrt{v_x^2 + v_y^2}$ their population tends to a χ distribution (a distribution that is the square root of the sum of the squares of two variables following normal laws). Finally, x, y components of numbers obtained after many collisions reach the maximum values of Shannon entropy (and the minimum of Boltzmann's H function).

The following table reports from [28] the rule of collisions, called **Pythagorean recombination game**. Diagrams are also reported in [28], at several collisions, which show clearly the emergence of an evident χ-distributions and the minimum of H function (the last diagram of the experiment is given in Fig. 5.2). In conclusion, according to Boltzmann's claim, the time arrow, that is, time-orientedness, is a consequence of the collective behaviour of a huge of elementary events following newtonian mechanics with no time orientation. Synthetically, time is a statistical phenomenon of complex systems (Fig. 5.1 and Table 5.2).

Table 5.2 The Pythagorean recombination game

Randomly choose two numbers a, b of the given number population;
Randomly choose a number a_1, such $a_1 \leq a$, and split a into a_1 and $a_2 = \sqrt{a^2 - a_1^2}$;
Randomly choose a number b_1, such $b_1 \leq b$, and split b into b_1 and $b_2 = \sqrt{b^2 - b_1^2}$;
Replace the pair a, b with the pair $a' = \sqrt{a_1^2 + b_2^2}$, $b' = \sqrt{b_1^2 + a_2^2}$.

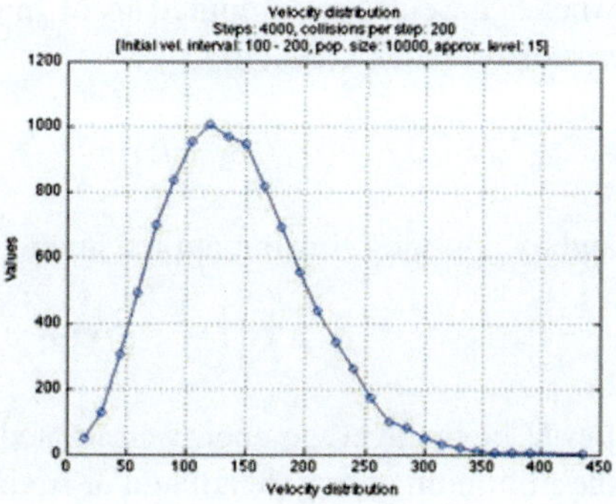

Fig. 5.2 4000 steps with 200 collisions per step, applied to the initial random distribution of velocities given in Fig. 5.1, with a specific approximation level, computed by a MathLab function (from [28])

Planck's Quantum Discovery

On December 14th of 1900, Max Planck gave an epochal conference where he presented a solution to a physical problem showing a theoretical limitation of classical physics: the explanation of the observed distribution of irradiation frequencies with respect to the temperature of the irradiating body (usually referred as the black body irradiation problem). The solution presented by Planck was based on the assumption that energy can be exchanged only in discrete quantities called quanta. The experimental formula giving the average energy U_ν irradiated at frequency ν and at temperature T is the following, but all the attempts to explain this formula with classical arguments had been unsuccessful:

$$U_\nu = \frac{h\nu}{e^{\frac{h\nu}{kT}} - 1} \tag{5.33}$$

The starting point of Planck's argument for deriving the formula above was the notion of entropy S as developed by Ludwig Boltzmann, in terms of mechanical statistics, as the logarithm of the number W of microscopical configurations providing the observed macroscopic state:

$$S = k \lg W. \tag{5.34}$$

Let us consider a body that we assume as a population of N microscopical elementary irradiating bodies and let P be the number of elementary irradiators having a given frequency ν. We determine the number of possible configurations giving the same amount of energy irradiated at the given frequency. This number corresponds to the following combinatorial value:

$$\frac{(N+P-1)!}{N!(P-1)!} \approx \frac{(N+P)!}{N!P!} \tag{5.35}$$

whence the entropy computed according to Boltzmann's formula yields for a constant k, called Planck constant:

$$S = k\left[\lg(N+P)! - \lg N! - \lg P!\right] \tag{5.36}$$

and by applying Stirling approximation:

$$S = k\left[(N+P)\lg(N+P) - N\lg N - P\lg P\right]. \tag{5.37}$$

Let U be the average energy irradiated by the N elementary irradiators, then if ε is the minimum energy irradiated at frequency ν we have:

$$NU_\nu = P\varepsilon \tag{5.38}$$

or, dropping the subscript ν in the passages for the sake of simplification:

$$P = \frac{NU}{\varepsilon} \tag{5.39}$$

when we replace the value of P of the equation above in Stirling approximation (5.37) we obtain:

$$S = k\left[\left(N + \frac{NU}{\varepsilon}\right)\lg\left(N + \frac{NU}{\varepsilon}\right) - \frac{NU}{\varepsilon}\lg\frac{NU}{\varepsilon} - N\lg N\right] = \tag{5.40}$$

$$S = k\left[N\left(1 + \frac{U}{\varepsilon}\right)\lg N\left(1 + \frac{U}{\varepsilon}\right) - \frac{NU}{\varepsilon}\lg\frac{NU}{\varepsilon} - N\lg N\right]. \tag{5.41}$$

Let us derive with respect to $U_{tot} = NU$. We observe that the derivative of a function $y\log y$, where y' the derivative of y, is $y'\log y + y'$, and that derivatives with respect to U of $N(\frac{1+U}{\varepsilon})$ and $\frac{NU}{\varepsilon}$ are equal, therefore:

$$\frac{dS}{dU_{tot}} = \frac{dS}{NdU} = \frac{k}{N}\left[\frac{N}{\varepsilon}\lg N\left(1 + \frac{U}{\varepsilon}\right) - \frac{N}{\varepsilon}\lg\frac{NU}{\varepsilon}\right] = \tag{5.42}$$

$$\frac{k}{\varepsilon}\left[\lg N\left(1 + \frac{U}{\varepsilon}\right) - \lg\frac{NU}{\varepsilon}\right] = \tag{5.43}$$

but $\frac{dS}{dU_{tot}} = \frac{1}{T}$, therefore:

$$\frac{dS}{dU_{tot}} = \frac{k}{\varepsilon}\left[\lg N\left(1 + \frac{U}{\varepsilon}\right) - \lg\frac{NU}{\varepsilon}\right] = \frac{1}{T} \tag{5.44}$$

that is:

$$\lg \frac{N(1+\frac{U}{\varepsilon})}{\frac{NU}{\varepsilon}} = \frac{\varepsilon}{kT} \tag{5.45}$$

whence:

$$\frac{N(1+\frac{U}{\varepsilon})}{\frac{NU}{\varepsilon}} = e^{\frac{\varepsilon}{kT}} \tag{5.46}$$

that, with the following simple passages, provides a final equation:

$$\frac{(1+\frac{U}{\varepsilon})}{\frac{U}{\varepsilon}} = e^{\frac{\varepsilon}{kT}} \tag{5.47}$$

$$\frac{\varepsilon}{U} = e^{\frac{\varepsilon}{kT}} - 1 \tag{5.48}$$

$$U = \frac{\varepsilon}{e^{\frac{\varepsilon}{kT}} - 1} \tag{5.49}$$

now setting $\varepsilon = h\nu$ as the energy quantum we obtain a formula coinciding with the experimental formula of irradiation given at beginning of this analysis, where subscript ν is explicit:

$$U_\nu = \frac{h\nu}{e^{\frac{h\nu}{kT}} - 1}. \tag{5.50}$$

Schrödinger Equation

In 1926 the physicist Erwin Schrödinger published a paper [23] where he presented an equation describing the state ψ of a quantum system, by assuming that it is described by a wave. The derivation is based on a particle of mass m and velocity v, to which, according to de Broglie's complementarity principle, a wave is associated with length h/mv, where h is the Planck constant, and for which the wave energy E verifies the Planck equation $E = h\nu$, where ν is the wave frequency.

The following are the main passages of Schrödinger's equation.

A wave of length λ and frequency ν has a velocity u given by:

$$u = \lambda\nu \tag{5.51}$$

and D'Alembert wave equation states that:

$$\Delta\psi - \ddot{\psi}/u^2 = 0 \tag{5.52}$$

Moreover, according to Planck, the wave frequency ν gives a quantum of energy E divided by the Planck constant h:

$$\nu = E/h \tag{5.53}$$

Equations (5.51) and (5.53) provide:

$$u = \lambda E/h \tag{5.54}$$

moreover, according to De Brogile, a wave length λ is associated to a particle of mass m moving at velocity v that satisfies the following equation:

$$\lambda = h/mv \tag{5.55}$$

and from Eqs. (5.54) and (5.55) we obtain:

$$u = E/mv \tag{5.56}$$

but considering the total energy E of a particle at velocity v, if the particle is subjected to a conservative field with a potential V, then E is the sum of kinetic energy and of potential energy V, that is, $E = \frac{1}{2}mv^2 + V$, thus we can write:

$$mv = [2m(E - V)]^{\frac{1}{2}} \tag{5.57}$$

and putting together Equations (5.56) and (5.57) we obtain:

$$u^2 = E^2/2m(E - V) \tag{5.58}$$

where we remark that the velocity v of a particle is different from the velocity u associated of its corresponding wave.

Schrödinger assumes for ψ a typical form of wave function (k a constant):

$$\psi = e^{-i(\omega t + k)} \tag{5.59}$$

where $\omega = 2\pi\nu$. But, according to Planck, $\nu = E/h$ therefore $\omega = 2\pi E/h$ thus, by deriving twice with respect to t we have:

$$\ddot{\psi} = -4\pi^2 E^2 \psi/h^2. \tag{5.60}$$

By replacing the right member of Equations (5.60) into (5.52) where u^2 is replaced according to Equation (5.58) we get exactly the original form of Schrödinger Equation (5.16) in his paper of 1926:

$$\Delta\psi + 8\pi^2 m(E - V)\psi/h^2 = 0 \tag{5.61}$$

now by setting $\hbar = h/2\pi$ and by multiplying by $\hbar^2/2m$ Equation (5.61) we have:

$$\hbar^2 \Delta\psi/2m + (E - V)\psi = 0 \tag{5.62}$$

or

$$-\hbar^2 \Delta\psi/2m + V\psi = E\psi \tag{5.63}$$

which is the standard form of Schrödinger Equation without time dependency.

If we introduce the Hamiltonian operator $\mathscr{H}$ (an analogous of that one of classical mechanics) such that:

$$\mathscr{H}(\psi) = -\frac{\hbar^2}{2m}\Delta\psi + V\psi$$

then, we obtain the following very synthetic form:

$$\mathscr{H}(\psi) = E\psi. \tag{5.64}$$

In order to introduce explicit time dependency in the equation, we derive $\psi = e^{-i(\omega t+k)}$ with respect to t:

$$\frac{d\psi}{dt} = -i\omega\psi = -i2\pi\nu\psi = -i\frac{2\pi E}{h}\psi \tag{5.65}$$

whence:

$$E\psi = \frac{d\psi}{dt}\hbar i \tag{5.66}$$

thus, from Eq. (5.63) the time dependent Schrödinger Equation follows:

$$-\frac{\hbar^2}{2m}\Delta\psi + V\psi = \hbar i\frac{d\psi}{dt}. \tag{5.67}$$

or, more synthetically:

$$\mathscr{H}(\psi) = \hbar i\frac{d\psi}{dt}. \tag{5.68}$$

The formal derivation of Schrödinger Equation can be easily followed, but what really does it mean? What intuition does it express? This is a very difficult question, and it is very impressive that Schrödinger seems guided by a sort of unconscious intuition, which is a miraculous mixing of physical classical principles (d'Alembert) with new ideas introduced by quantum perspective (Planck's Energy-frequency relation and de Broglie's particle-wave complementarity).

First of all, what ψ is? Certainly, it is a wave, but what kind of wave? Surely, Schrödinger refers to a subatomic particle, but the implicit assumption was that the approach extends to systems of particles, up to a whole atom. These systems can be correctly described only by introducing new principles with respect to classical mechanics and electromagnetism. For example, the Hydrogen atom cannot be explained by using classical electrodynamics. In fact, an electron rotating around a proton, in its motion, should consume a quantity of energy that would produce a collapse of its rotation toward the proton.

Schrödinger's approach assumes that ψ describes the state of a quantum system. If a number of variables identify a quantum system, at any moment, these variables "oscillate", as a wave, within specific ranges of values, and each value is assumed with a given probability of occurring. This interpretation is not present in the 1926 original paper of Schrödinger, who believed in a real physical wave (he uses the term "parcel of waves"). Namely, the probabilistic interpretation emerged later, especially for the contributions of Max Born. In this sense, a state is a system of probability distributions, that is, a system of information sources, in the sense of Shannon.

The state ψ of a quantum system is a wave, but a "probability wave", that is, an abstract wave, very different from the usual ones that propagate through a medium (the empty space in the case of electromagnetic waves). We cannot observe ψ without disturbing it, and a datum is generated as the output of an observation that contemporaneously changes the original setting of the observed state.

This mechanism is strictly regulated by a mathematical theory initiated by John von Neumann, based on Hilbert spaces over complex numbers, with several kinds of operators acting on these spaces. Of course, the story is much more complex and the interested reader can find many beautiful presentations of quantum principles in many classical and new presentations of quantum mechanics [29–32].

What is really surprising is the fact that Schrödinger Equation, which is the kernel of quantum mechanics, captures the internal structure of a quantum state. However, it was discovered by Schrödinger by means of passages where ψ is considered as a physical wave, by applying classical physical principles integrated with Planck's $E = h\nu$, and de Broglie's $\lambda = h/m\nu$. The situation is resembling Colombo's discovery of America, while he was searching for a way to India. In this case, the explanation is due to the ignorance about the existence of a country between the Atlantic and Pacific oceans. In the case of Schrödinger's discovery, no clear explanation is available up to now. This fact suggests us the presence of a mystery strictly related to the role of information in physical models. In other words, we can read Schrödinger's derivation as a sort of "unreasonable passage" from a physical reality to an "informational essence", hence a crucial open problem remains for the explanation of this passage.

Information results to be a sort of red thread of the most important physical revolutions of the last centuries: Boltzmann's entropy is an analogue of Shannon entropy; quanta emerge from a discretization of irradiating oscillators, by considering Boltzmann's entropy of the whole oscillating system; finally, quantum systems are information sources of physical variables.

This analysis shows as the informational perspective involves, mathematics, physics, biology, and of course neurology, being neurons, of natural or artificial systems, computing agents integrated into networks and communicating with signals of continuous and/or discrete nature. All the levels of reality, and of our knowledge of reality, are dominated by information sources, and many secrets of this reality can be deciphered by deep analyses of their informational mechanisms.

Since Boltzmann's discovery of the H function, an intrinsic relationship emerges between mass-energy and data-information. Symbols need physical support in order to be realized, therefore information postulates physics. Conversely, from quantum physics, we discover that physics requires information sources for expressing the physical states of microscopic quantum systems. This means that the physical and informational perspectives seem to be different aspects of the same reality. The bounds and the passages from these two faces of the same coin are again the missing points that the science of the future is called to analyze and understand. This theoretical necessity has been raised by many important physicists, along different lines, by reconsidering physical principles of quantum mechanics [25, 33–37], and a famous physicist, as John Archibald Wheeler, in his paper "Information, physics, quantum: The search for links" [25] writes: [...] *Every physical quantity, every it, derives its ultimate significance from bits, binary yes-no indications, a conclusion which we epitomize in the phrase* **it from bit**.

We are waiting for the right theoretical framework for being this link adequately developed.

In conclusion, it is worthwhile to mention here the famous Schrödinger's booklet [38] entitled *What is Life?* published in 1944. The great physicist, following a very general reasoning, prophesies the presence, inside cells, of some kind of structure encoding biological information. Only a few years before the DNA discovery, he speaks of "aperiodical crystals". Thus, the scientist who introduced the quantum wave equation postulates a chemical structure supporting biological information. Science is very often compartmentalized in disciplines, but deep concepts do not follow the academic partitions of scientific fields.

5.4 The Informational Nature of Quantum Mechanics

Let us briefly outline the principles of quantum mechanics (QM), in a way that puts in evidence the crucial role of information as a founding concept.

QM, for its nature, is on the border of knowledge of physical reality. However, it seems that the lack of a rigorous and coherent explanation of its principles, at the same time firmly anchored to physical intuitions, is mostly due to an inadequacy of language. In fact, great difficulties arise when classical concepts, which were elaborated for macroscopic phenomena, are used in QM. The mental images evoked by terms such as particle, wave, measure, ... become a barrier to the comprehension of microscopic realities. In the following, we give a formulation, which surely requires further speculation, where opposite terms are conciliated by distinguishing between **epiphenomenon** and **ipophenomenon**, a sort of reinterpretation of *hidden variables*, often invoked in the long debate on QM. Just this distinction allows us to introduce the notion of **localization wave**, which could make possible to link the original physical intuition of Schrödinger's wave with Born's probabilistic interpretation.

Quantum Phenomenon

A quantum phenomenon consists of the interaction between two systems Q and O. The first one is a quantum system, that is, a microscopic system at the atomic or subatomic level (one or more particles, an atom, an electron, a photon beam, a laser ray, ...). The second one is a macroscopic observation instrument (a Geiger counter, a Stern-Gerlach apparatus, a polarization filter, ...). A microscopic system cannot be observed without an observation instrument, that being a macroscopic object, follows a logic assuming the principles of classical mechanics and electromagnetism. Therefore, even if the laws of microscopic systems are different from those of classical physics, an observation assumes these laws.

Epiphenomenon, Ipophenomenon

A quantum system Q is characterized by some physical variables divided into two disjoint groups. The first group are the internal variables of the system, the remaining are the variables observed during the interaction with the observation system O. The observed variables assume, after the interaction, values that determine the epiphenomenon of the interaction. The values of the internal variables determine the ipophenomenon of the interaction. In particular, at any state s of Q, a complex number $\psi(s)$ is associated, called **probability amplitude** of s. This amplitude is not observed, but it is linked to some observable value $q(s)$ by the fundamental relation:

$$Prob(q(s)) = |\psi(s)|^2$$

according to which the probability of observing $q(s)$ is given by the square of the absolute values of $\psi(s)$.

Localization Wave

A quantum particle (photon, electron, proton, neutron, neutrin, positron, ...) is the space localization of a quantum wave, and conversely, a quantum wave is the potentiality of space localization of a particle. This means that the probability wave usually associated with Schrödinger equation, according to Born's interpretation, corresponds to a true physical wave, a space vibration, which we know as a probability wave. In other words, Planck's principle claiming that energy is present in packets $h\nu$, with h Planck constant and ν a frequency of oscillations, has to be properly intended so that energy can be transferred between systems in discrete packets. Then, in this perspective, a particle is the quantum packet of a wave that transfers its energy, in a specific **microregion**, to an observation instrument, revealing the interaction between the wave and the instrument. Therefore Schrödinger wave and Born probability wave are the two faces of the same coin. This interpretation can explain something incredible: the fact that Schrödinger, starting from a physical intuition, expressed by d'Alembert wave equation, and inserting in that equation Planck and de Broglie equations, by using principles of classical mechanics, was able to discover his famous ψ-equation, later interpreted as a probability wave. The duality wave-particle is the essence of quantum phenomena, where the wave is the ipophenomenon related to a particle that is the epiphenomenon revealing it.

Another aspect of a localization wave is the intrinsic random nature of its vibration around positions or centres of localization. This randomness is the basis of randomness observed in quantum phenomena, resembling random walks, and which constitutes the basis for true random variables based on physical phenomena (observed in Brownian motions, or in the arrival time of photons in detectors).

Many books on QM start from the interference phenomenon of Young revealing the wave nature of light, against Newtonian particle viewpoint about light rays. In 1801, Young presented a famous paper to the Royal Society entitled "On the Theory of Light and Colours'. In Young's experiment, light passes through two slits and reaches a screen, where light appears divided by dark stripes, which become wider as they are more remote from the slits. This phenomenon is completely explained by assuming that light consists of waves, which are subjected to **interference** when they meet in opposite phases that destroy each other by producing darkness.

The particle nature of light was again accredited after the discovery of the photoelectric phenomenon, exhibiting a clear particle behaviour of photons, seen as light quanta.

However, reconsidering Young's experiment, where single photons are sent to two slits, for example, one photon per second, the same interference is obtained, while interference disappears (no stripes, but only two light regions on the screen, more luminous in correspondence to the slits) when an observation apparatus is located near to one of the two slits, for knowing through which one any photon will pass.

This situation seems paradoxical because it results that photons behave in completely different ways when they are observed, with respect to when they are not observed. Moreover, the same situation arises even when electrons, instead of photons, are considered. Wave interference is the core of QM, where the dualism between waves and particle is a crucial aspect of quantum phenomena.

Non determinism

Laws that describe quantum phenomena, differently from those of classical physics, do not tell us what happens to a system when it is in a given state. Quantum laws tell us what is the probability of observing one value of an observed variable X, among some possible values that the variable can assume., If q_0 is the state of the quantum system that we observe and $a_1, a_2, \ldots a_j, \ldots$ are the possible values of the observed variable X, then what we can know from a quantum description are the probabilities $p_1, p_2, \ldots, p_j, \ldots$ of observing the corresponding values of X. If $q_0(X)$ is the value of X observed in the sate q_0, we can write:

$$p_i = Prob(q_0(X) = a_i).$$

this means that a quantum description provides a Shannon source of information associated with the observed state.

Let us consider a short presentation of some quantum principles, in light of the concepts previously outlined. The presented perspective is aimed at giving possible solutions to two crucial conundrums of QM: the wave interference experiment and Schrödinger's equation derived from the intuition of a physical wave, but interpreted as an abstract probability wave. Here, the wave-particle dualism is based on the epiphenomenon-ipophenomenon distinction and on the concept of localization wave.

1. **Discreteness** (Planck) 1900
Energy is exchanged among physical systems in discrete packets. $E = h\nu$ (h Planck constant, minimum action quantum, of order 10^{-34} Joule x sec, ν wave frequency).
2. **Duality** (de Broglie) 1923
A wave of length h/mv is associated with any particle of mass m and velocity v. This principle is the kernel of all the following principles and, in particular, the basis of Schrödinher's equation, which is the epicentre of QM.
3. **Exclusion** (Pauli) 1925
Two identical particles cannot be in the same quantum state. This simple property allows for explaining the logic of Mendelejev's table of chemical elements, and many important properties of the different forms that matter assumes in the universe.
4. **Superposition** (Schrödinger) 1926
An event is given by the assumption of values from the variables defining the state of a quantum system. We write $x_1 = a_1,\ x_2 = a_2, \ldots$ for denoting an **event** of variables $x_1, x_2, \ldots$. An event can be defined apart from any possibility of observing the values of the considered variables. For example, x_1 could be an internal variable and then a_1 is never observable, while an **experiment** is an event where the values of some variables are observed.
A quantum system is described by a periodical function ψ (wave) taking values in the complex plane, or in general, in vectors of complex numbers. The value $\psi(s)$ for a state s of the system Q is called **wave amplitude** associated with the state s. This function verifies the following Schrödinger's equation:
$$\mathcal{H}[\psi(s)] = E\psi(s)$$
according to which the Hamiltonian operator $\mathcal{H}$ applied to $\psi(s)$ provides the value $\psi(s)$ multiplied by the value E of the energy of the system, where the Hamiltonian $\mathcal{H}$ generalizes an analogous operator of classical mechanics. In the ψ-equation given above, three variables occur: the state variable s, the energy variable E, and the amplitude variable $\psi(s)$. We easily realize that two of them are non-observable variables (s, $\psi(s)$), while E is an observable variable. The equation expresses an intrinsic connection among them, which is the fundamental relation of quantum systems.
5. **Indetermination** (Heisenberg) 1927
$$\Delta x \Delta p \geq h$$

where, for any variable y, notation Δy denotes the approximation of determination, that is the error interval in the determination of y. If the product $\Delta x \Delta p \geq h$, a small error in determining the position x of a particle corresponds to a very large error in determining its moment p, and *vice versa*.

6. **Complementarity** (Bohr) 1927
The description of a quantum phenomenon postulates an interaction between a quantum system Q and a macroscopic observation system O (instrument). These two physical systems are complementary and their interaction produces a perturbation of the observed system, which in its essence remains unknowable. Therefore, if we observe a quantum system, then we disturb it. At the same time, we cannot observe it without disturbing it. This means that the description of a quantum system is a description of its perturbations. This principle is the basis of other specific complementary aspects that are essential in QM.
7. **Probability** (Born) 1927
The probability of realizing a state s is given by the square of $|\psi(s)|$, the absolute value of $\psi(s)$:
$$Prob(s) = |\psi(s)|^2.$$

If an experiment can discriminate a value of an observed variable x among many different possible values, then the probability of the experiment is the sum of the probabilities of all single possible alternatives. For example, if x is an observed variable that can assume only two possible values a_1, a_2, then:

$$Prob(x = a_1 \ \ or \ \ x = a_2) = Prob(s(x) = a_1) + Prob(s(x) = a_2).$$

Conversely, let x be a variable that is not observed, and either $s(x) = a_1$ or $s(x) = a_2$, then, according to superposition:

$$\psi(s) = \psi(s(x) = a_1) + \psi(s(x) = a_2)$$

therefore:

$$Prob(s) = |\psi(s(x) = a_1) + \psi(s(x) = a_2)|^2$$

but the square of a sum is different from the sum of squares, this means that there is **interference** between the two alternatives. This situation is that one of Young's interference experiments revealing the wave nature of light (Wave mechanics is another way of denoting Shrödinger's approach). The amplitude/probability interference is a crucial aspect in discriminating between observed and non-observed variables. Namely, an event involving variables that are not observed follows the rule of summing amplitudes, which produces superposition, whereas, in an experiment observing variables the probability of more alternatives is the sum of the single probabilities.

8. **Representation** (Dirac-von Neumann) 1925–1932
A quantum state is a vector in a multidimensional Hilbert space over complex numbers satisfying Dirac-von Neumann's Axioms (on the scalar product, orthogonality, linear, hermitian, and unitary operators).
9. **Pilot Wave** (Bohm-deBroglie) 1952–1957
The probability wave ψ corresponds to a physical wave (that in the case of a particle guides its movement). This principle is a natural consequence of the wave approach in QM, and corresponds to the "localization wave" of the present formulation. However, it requires further theoretical and experimental developments. Moreover, the relationship between the localization wave and probability wave is linked to the role that randomness plays in QM. This aspect is surely a key to understanding the informational nature of QM. This research line is probably related to other fields of physical research (String Theory, Quantum Gravity), where the submicroscopic structure of space (called *plenum* by David Bohm) could reveal fundamental informational aspects of physical reality, summarized by the famous John Wheeler's motto *it from bit*.
10. **Non-Locality** (Entanglement) (Bell) 1964–1982
The principle of locality in physics establishes that the state of a system depends only from what is spatially "close" to it. In 1964, the British physicist John Bell, developed an analysis, originating from a version of EPR paradox due to David Bohm. This paradox was formulated by Einstein, Podolsky, Rosen for showing that QM has something wrong or incomplete. Bell found an inequality that has to be satisfied by any physical theory that is in agreement with the locality principle. From 1972 up to 1982, the physicists Alain Aspect, John F. Clauser, and Anton Zeilinger conducted experiments confirming that Bell's inequality does not hold in some quantum systems. In these systems, it is possible that two particles, even at a very large distance, behave as if they were a unique reality, because any change in one of them, instantaneously, produces a corresponding change in the other one. This phenomenon called **entanglement** is a specific feature of QM. In 2023 the physicists who experimentally proved quantum non-locality received the Nobel prize for Physics. Namely, quantum entanglement represents a frontier for theoretical and applicative developments of QM (quantum computing and quantum cryptography).

References

1. Feller, W.: An Introduction to Probability Theory and Its Applications. Wiley, New York (1968)
2. Bonnici, V., Manca, V.: An informational test for random finite strings. Entropy **20**(12), 934 (2018)
3. NIST, National Institute of Standards and Technologies. A Statistical Test Suite for Random and Pseudorandom Number Generators for Cryptographic Applications. Gaithersburg, MD 20899-8930, Revision 1a (2010)

4. Hastings, W.K.: Monte Carlo sampling methods using Markov chains and their applications. Biometrika **57**(1), 97–109 (1970)
5. Havil, J.: Gamma: Exploring Euler's Constant. Princeton University Press, Princeton (2003)
6. Brin, M., Stuck, G.: Introduction to Dynamical Systems. Cambridge University Press, Cambridge (2002)
7. Bonnici, V., Manca, V.: Informational laws of genome structures. Sci. Rep. **6**, 28840. http://www.nature.com/articles/srep28840 (2016) Updated in February 2023
8. Bonnici, V., Manca, V.: Author correction: informational laws of genome structures. Sci. Rep. **13**, 3422 (2023)
9. Chaitin G.J.: Algorithmic Information Theory. Cambridge University Press, Cambridge (1987)
10. Kolmogorov, A.N.: Three approaches to the quantitative definition of information. Probl. Inf. Transm. **1**, 1–7 (1965)
11. Martin-Löf, P.: The definition of random sequences. Inf. Control **9**, 602–619 (1966)
12. Borel, E.: Les probabilities denomerable et leurs applications arithmetiques. Rend. Circ. Mat. Palermo **27**, 247–271 (1909)
13. Borel, E.: *Leçons sur la théorie des functions*. Gauthier-Villars (1914)
14. L'Ecuyer, P.: History of uniform random number generation. In: Proceedings of the 2017 Winter Simulation Conference, pp. 202–230 (2017)
15. L'Ecuyer, P.: Random number generation with multiple streams for sequential and parallel computers. In: Proceedings of the 2015 Winter Simulation Conference, pp. 31–44. IEEE Press (2015)
16. Marsaglia, G.: DIEHARD: A Battery of Tests of Randomness (1996). http://stat.fsuedu/~geo/diehard.html
17. Nielsen, M.A., Chuang, I.L.: Quantum Computation and Quantum Information. Cambridge University Press, Cambridge (2000)
18. Nies, A.: Computability and Randomness. Oxford University Press, Oxford (2009)
19. Volkenstein, M.V.: Entropy and Information. Springer, Berlin (2009)
20. Brush, S.G., Hall, N.S. (eds.): The Kinetic Theory of Gases. An Anthology of Classical Papers with Historical Commentary. Imperial College Press, London (2003)
21. Sharp, K., Matschinsky, F.: Translation of Ludwig Boltzmann's Paper "On the relationship between the second fundamental theorem of the mechanical theory of heat and probability calculations regarding the conditions for thermal equilibrium." Entropy **17**, 1971–2009 (2015)
22. Planck, M.: Planck's Original Papers in Quantum Physics. Annotated by Kangro, H., translated by Haarter D., Brush, S.G. Taylor & Francis, London (1972)
23. Schrödinger, E.: An undulatory theory of the mechanics of atoms and molecules. Phys. Rev. **28**, 6 (1926)
24. Wiener, N.: Cybernetics or Control and Communication in the Animal and the Machine. Hermann, Paris (1948)
25. Wheeler, J.A.: Information, physics, quantum: the search for links. In: Zurek, W.H. (ed.) Complexity, Entropy, and the Physics of Information. Addison-Wesley, Redwood City, California (1990)
26. Carnot, S.: Reflections on the motive power of heat (English translation from French edition of 1824, with introduction by Lord Kelvin). Wiley, New York (1890)
27. Manca, V.: An informational proof of H-Theorem. Open Access Lib. (Modern Physics) **4**, e3396 (2017)
28. Manca, V.: Infobiotics: Information in Biotic Systems. Springer, Berlin (2013)
29. Bohm, D.: *Quantum Theory*. Dover (1989)
30. Dirac, P.: The Principles of Quantum Mechanics. Oxford University Press, Oxford (1958)
31. Susskind, L., Friedman, A.: Quantum Mechanics. Basic Books, The Theoretical Minimum (2014)
32. Lederman, L.M.: Quantum Physics for Poets. Prometheus Books (2011)
33. Bohm, D.: Causality and Chance in Modern Physics. Routledge & Kegan Paul, London (1957)
34. Bell, J.S.: Speakable and Unspeakable in Quantum Mechanics. Collected Papers on Quantum Philosophy. Cambridge University Press, Cambridge (2004)

35. Fagin, F.: Consciousness comes first. In: Consciousness Unbound. Rowman& Littlefield, Lanham (MD) (2021)
36. D'Ariano, G.M., Perinotti, P.: Derivation of the dirac equation from principles of information processing. Phys. Rev. A 062106 (2014)
37. D'Ariano, G.M., Chiribella, G., Perinotti, P.: Quantum Theory from First Principles. An Informational Approach. Cambridge University Press, Cambridge (2017)
38. Schrödinger, E., What Is Life? the Physical Aspect of the Living Cell and Mind. Cambridge University Press, Cambridge (1944)

Chapter 6
Life Intelligence

Introduction

Let us introduce the theme of the relationship between life and intelligence by reporting the recent success of AI (Artificial Intelligence). Protein folding is one of the most challenging problems of biology. In fact, when a protein is synthesized from an RNA transcript to a sequence of amino acids, according to the genetic code defined over codons (by means of ribosomes and RNA-transfer elements), in a very short time, it assumes a spatial form due to many kinds of repulsion and affinity forces between the different amino-acids of the sequence. This form determines the protein function.

Guessing the form associated with a given sequence of amino acids is a very complex task that has remained unsolved for decades. Recently, a solution was found, which was based on Alpha-go, a machine learning program used in designing a machine able to play go, at a level of competence able to win against a world champion. The same program, suitably instructed along years of training, solved the protein folding, by obtaining an epochal success of artificial intelligence. In this way, science solved the problem, but without knowing exactly how the solution was obtained. This is a peculiar aspect of human intelligence, giving us behavioural competencies, acquired by training, but often confined at the level of our unconscious neural activity, such that we are able to do something, but we do not know how we do it. In the same manner, an AI algorithm is something internalised in a structure, as a path in a suitable space, directing from a target toward a corresponding solution, where it is unknown the relationship between the path and the solution and between the training and the emergence of the path. However, we know the meta-algorithm, that is, the rules that allow a suitable structure to evolve, along the training, toward a configuration where the solving path can emerge.

Let us give a preliminary approximate definition of intelligence. This is, of course, a philosophical question but it has a huge impact on defining *intelligent* theoretical

V. Manca and V. Bonnici, *Infogenomics*, Emergence, Complexity and Computation 48,
https://doi.org/10.1007/978-3-031-44501-9_6

frameworks and on implementing them. A past definition of intelligence was related to the emulation, or similarity, of human activities. An entity was considered intelligent if it was able to act as a human brain. It is today clear that human intelligence is not the only form of intelligence on our planet. Many other animals show similar forms of intelligence. In animals, intelligence is often a property of an individual which is associated with the brain, and thus with neurons. For this reason, many modern computational models of intelligence try to emulate the brain. In fact, artificial neural networks (ANN) are inspired by animal brains, and in particular, they try to emulate the complex network of neurons of which a brain is composed. However, many different forms of intelligence exist which do not (as far as our knowledge) rely on neurons. Plants are intelligent organisms that interact with the environment and are also able to communicate, but they do not have neural cells. Similarly, microbes are unicellular organisms showing the same capabilities, even if they are made of a single cell. Thus, intelligence is something more abstract that the results of the connections of neurons. A modern definition of intelligence defines a (complex) system as intelligent if such a system is able to: (i) retrieve information from the environment; (ii) solve a specific task/problem by building a strategy by exploiting the retrieved information; (iii) store the information and the strategy in order to face up to the same problem in future times, promptly; (iv) exploit the learned strategy to solve similar problems. An intelligent system may be formed by more than one entity, such that the entities interact to obtain the wanted capabilities.

All forms of intelligence present some fundamental aspects that are independent of their specific realizations. Typical structures of intelligent systems are networks, as in animal nervous systems consisting of neurons and synapses among them. However, substances connected by biochemical transformations of molecules (reactants) into other molecules (reagents) provide other kinds of networks. Analogously, genes connected among them and with proteins (factors and co-factors) by activation and inhibition relations provide other very rich and sophisticated networks. In this sense, life is based on networks, starting from the metabolic level of biochemicals, up to the genetic and cellular levels, and beyond them, with species and ecosystem networks. All these systems exhibit very big connectivity, which can be defined as hyper-connectivity. Moreover, very often "intelligent" structures are hyper-heterogeneous (with many different categories of components), hyper-dimensional (dependent on several parameters), hyper-stratified and hyper-modular (organised in functional units). Finally, random events very often affect their dynamics.

For what concerns modern computational models, emerging in the context of artificial intelligence, the retrieving of specific information from the environment is today a mostly solved problem. There are plenty of sensor technologies that scan the environment and produce digital data. Starting from the acquired data, the creation of data-driven strategies and their exploitation is the main focus of the so-called machine learning models, which have shown incredible performance in recent years. In particular, modern ANNs are very good at solving problems of classifying objects, namely, assigning an object to a specific predefined class by observing the features of such an object. These solutions postulate that a problem can be solved by learning the parameters of a specific computational model. The learning process is based on a learn-by-example approach in which example data are used to learn the parameters. This implies that examples must be available before the learning phase and, often, the number of available examples affects the performance of the learned model. In fact, in general, the more input example is available the higher the performance of the model is. Unfortunately, it is very uncommon to observe machines with flexible and wide-spectrum learning strategies. In fact, it is usual that models are trained on data that regard a specific problem, thus they become able to solve just that specific problem. It has to be noticed that, instead, the capability of re-purposing a strategy developed for a different problem to another problem, as well as the reusing of information previously acquired for a different task, remains a capability peculiar to living organisms.

Organisms aggregate to form communities, where particular forms of intelligence arise from such aggregation. In this context, swarm intelligence and, more in general, forms of collective intelligence are well-known to scientists and are sometimes used as inspiration for computational models. At the bottom level of life intelligence, we may find genomes. They are surely the directors of living cells, thus they may seem intelligent entities, but something more interesting regards them. In fact, if we see genomes not as single entities but as a population, which evolves over time, then a different kind of intelligence emerges. This intelligence differs from collective intelligence because the entities that are interested do not collaborate in an explicit way, and often they do not belong to the same temporal frame. Thus, the intelligent genomic system is not a single genome, nor a single lineage of genomes, but it is the set of genomes which has evolved to adapt to a given environment. In this perspective, genomes constantly interact with the environment, which defines the constraints for their survival. They acquire information by sensing the environment, but this is not the primary form with which they learn. In fact, their learning process is more related to a try-and-survive-or-die approach. They made *random* choices which lead to characteristics of the organisms they implement which advantageous or disadvantageous property is determined *a posteriori* by the environment. However, such choices are not uniformly (purely) random, but they take into account the information that the system has previously acquired. Once a genome learns that a given choice is advantageous, it tries to keep such a choice for future generations and to protect it. In this perspective, genomes have developed strategies to recover from DNA reading errors and to reduce the variation level of certain regions of the DNA molecules. Once a certain choice, namely, a variation in the genomic sequence, has been made, it is used to implement a potential set of functions of the living cell/organism. Thus, the function that appears to be advantageous is only one with such a set of potentialities. This behaviour gives genomes, and more in general to all biological systems, the capability of being able to solve new previously-unseen problems by exploiting solutions to old already-seen problems. Scientists had exploited the knowledge regarding genomes and biological systems to produce computational models for solving problems, which are briefly described below in this chapter. A very general aspect emerges for this knowledge, that is the exploitation of "informed" random variables. Such a random engine is used to make choices that bring to potentially unexpected functions, but, at the same time, it is built by a learning process which preserves advantageous past choices. Such advantageous choices represent a piece of information that the system acquires by testing and sensing the environment.

Two important aspects seem to be necessary for the emergence of intelligent behaviours, within a given structure, as it happens in biological networks: (1) mechanisms of memorisation, (2) mechanisms of reflexivity, where some parts become able to control the whole structure, showing a sort of conscience. Genomes, probably, emerged as sorts of memories of complex enzymatic activities, responsible for cell competencies. Moreover, any genome includes a kernel that is able to control the preservation and a coherent unfolding of biological knowledge accumulated along its evolution.

In what follows, we describe the three main types of bio-inspired computational models by putting a special focus on how they exploit random variables to reach their goals. The first two types, which are genetic algorithms and swam intelligence, are usually applied to solve combinatorial or optimisation problems. Instead, the third type, namely the artificial neural networks, is generally applied to solve classification problems, but they find applicability to other types of problems.

6.1 Genetic Algorithms

The resolution of combinatorial problems, and in particular constraint satisfaction problems, are within the range of applicability of genetic algorithms [1]. A constraint satisfaction problem (CSP) consists in finding a specific assignment of values to a given set of variables such that a predefined set of constraints is satisfied [2]. For each variable, a set of possible values is allowed, and different variables may share possible values. Because of their combinatorial nature, many CSP problems are NP-Complete problems and no polynomial algorithm exists for their resolution. A well-known CSP problem is the graph colouring problem in which we want to assign a colour to the vertices of a graph such that the same colour is never assigned to two neighbouring vertices [3]. Formally, a graph is a pair (V, E) such that $V = \{v_1, \ldots, v_n\}$ is the set of vertices and $E \in V \times V$ is the set of edges between such vertices. Given a set of colours C, the goal is to find a function $m : V \mapsto C$ such that $\forall(v_i, v_j) \in E \Rightarrow m(v_i) \neq m(v_j)$. The CSP problem is modelled such that V is the set of variables, C is the set of values and E defines the constraints to be verified.

A genetic algorithm aims at finding m by iterative steps which emulate the evolution of genomes and their natural selection due to environmental constraints. Each individual genome represents an assignment of values, one for each variable, which partially satisfies the set constraints. Thus a valid assignment m is obtained by evolving and reproducing a population of genomes. Thus, genomes are strings of length $|V|$ over the alphabet C. Each position i of a string relates to a specific vertex/variable v_i. Such a relationship is the same for every genome in the population. The steps of a genetic algorithm are initialisation, evaluation, selection, reproduction and mutation. Except for initialisation, the other steps are repeated in this exact order for a given number of iterations. The algorithm has two main parameters: the population size and the number of iterations. The algorithm starts by initialising a population $\mathbb{P} = \{G_1, G_2, \ldots, G_m\}$ of genomes. Initial genomes are usually created using a random process which picks up symbols in C with a uniform random distribution. However, more sophisticated rules can be used for their creation. For example, we may extract two different symbols in a single extraction and enqueue them to create the solution. Namely, we can extract 2-mers. Step 2 regards the evaluation of the current population with respect to their agreement to the constraints. A *fitness* function is defined such that the higher (lower) the value of the function for a given genome, the more suitable the genome is to solve the problem. For our example, we can define the fitness function as $\phi : \mathbb{P} \mapsto \mathbb{N}$ such that it represents the number of constraints (edges of the graph) which are violated. A genome G_x representing a valid complete solution has $\phi(G_x) = 0$, and the maximum value for ϕ is $|E|$. Thus, it is defined as $\phi(G_x) = |\{(i, j) : (v_i, v_j) \in E \text{ AND } G_x[i] = G_x[j]\}|$ with $1 < i, j < |G_x|, i \neq j$. However, the fitness function does not necessarily have to be defined directly on the constraints. For example, the fitness function may evaluate the number of positions of G_x (namely the number of vertices of the graph) which contribute to violating the constraints. In this case, the function is defined as $\phi(G_x) = |\{i : \exists 1 < j < |G||(v_i, v_j) \in E \text{ AND } G_X[i] = G_x[j]\}|$. Subsequently, a selection phase is run according to the evaluation results. The most suitable genomes are chosen to be reproduced, while less attractive genomes are discarded. Such a selection can be performed by ranking the genomes according to ϕ a by selecting the top-k individuals. However, such a procedure may drive the entire search procedure to get stuck in local minima which never lead to an optimal solution. For this reason, more sophisticated approaches allow non-top-k to be selected by defining specific rules that define the probability of an unattractive solution being selected. Local minima are critical issues for computational algorithms aiming at solving maximisation (minimisation). Algorithms, such as Tabu Search and simulated annealing, exploit randomness in a similar way to genetic algorithms do in the selection step. They allow apparently-unfeasible partial solutions to be selected with a (sometimes informed) random probability distribution. After the selection phase, a reproduction phase recombines the genetic material of survived genomes. A crossover operation between pairs of genomes is performed. Such a crossover can be made in several ways. A simple approach is to randomly pick up a position i of the genome and use such a position for a cut-and-paste operation. Thus, given two parent genomes, G_x and G_y, two new genomes, G_z and G_h, are generated such that $G_z = G_x[1, i] \cdot G_y[i + 1, n]$ and $G_h = G_y[1, i] \cdot G_x[i + 1, n]$, where n is the length of the genomes. Of course, such a reproduction process can be as sophisticated as wanted. More complex crossover operations can be defined, and uni-sexual or multi-sexual reproduction can act. Once the new progeny is created, it is added to $\mathbb{P}$. Because we usually want $\mathbb{P}$ to not increase in size, the parent genomes are removed from the population after their reproduction. Subsequently, the new genomes are mutated. The process was originally intended to emulate single nucleotide polymorphism. In general, it uses a uniform probability distribution to pick the position to be mutated and to pick a symbol to be assigned to such a position, but of course, it can be a more sophisticated process. Mutations allow the search process to visit a wider range of areas of the searching space, with respect to simply crossing the genomes. It is another powerful strategy for escaping from local minima. Evaluation, selection, reproduction and mutation are repeated until a solution is found.

It has to be noticed that there is an essential difference between how genomes actually act and how they are emulated to solve computational problems. In computational approaches, such as the one described above for the resolution of the graph colouring problem, constraints are constants, namely, they do not change over time. In real life, genomes have to face an environment that may mutate. For this reason, convergent evolution acts between species of different continents. However, local casual facts append in specific places, and genomes have to adapt to these short-term events. In conclusion, the evolutionary process partially fits a global fixed fitness function, thus it preserves past advantageous mutations. However, it is equipped to potentially react to temporary environmental changes which are faced by exploiting such an informed random process of adaptation.

6.2 Swarm Intelligence

Social behaviours within swarms of animals (birds, or insects) have been used as a source of inspiration to implement artificial intelligence. In a swarm intelligence algorithm, there are several entities, which correspond to elements of swarms, which simultaneously search for a solution. The algorithm has two main phases. An exploitative phase in which each entity, independently from the swarm, explores the proximal space in search of a better partial solution. Then, the information independently retrieved by each entity is shared with the whole swarm to combine such knowledge for driving the global search process in promising directions. Such approaches are then usually used in solving optimisation problems. One example is the pigeon-inspired optimisation (PIO) [4] which emulates strategies the pigeons use when flying as a swarm during migration. Each pigeon acquires information from the environment, for example, the sun's location, landmark markers and measurements of the magnetic field. Because such information is acquired individually by the pigeons, each individual has a personalised perspective of the environment which leads the pigeon to take a specific direction. However, other pigeons may have more useful data from the environment that can help the whole swarm to drive in a better trajectory. Moreover, a pigeon swarm is hierarchically organised into clusters, such that in each cluster a leader takes decisions and the other pigeons only share their information with the leader and follow the leader's decision.

Similar to a pigeon swarm, but without the hierarchical structure, there is the particle swarm optimisation algorithm (PSO) [5, 6]. A set of particles scan simultaneously the space in search of an optimal solution, for example, the minimum of a function with multiple parameters. Thus, each particle represents a position within the n-dimensional search space that is a specific parameters assignment. Moreover, each particle has a specific direction within the n-dimensional search space. At each step, the particles optimise their direction through an optimal (unknown) solution by taking into account the best position visited by the particle and the best position visited by any particle in the swarm. These two positions may produce contrasting directions, thus the resultant direction is a mediation of them by suitable weights. Particles are initially located in random (possibly equally spaced) positions of the search space, but this behaviour may be not enough to avoid the algorithm to trap in local minima. For this reason, often informed random choices are taken by the particles. Thus, the PSO algorithm is enhanced with techniques coming from the tabu search algorithm [7] or simulated annealing optimisation [8].

Is a living entity that always acts with the same pattern, by denying chances of random divergences, truly intelligent? Is informed randomness the key feature for intelligence, namely for adapting to a continuously changing environment?

6.3 Artificial Neural Networks

The first discrete mathematical model of nervous systems was Neural Networks (NN), defined in 1943 by Warren McCulloch (neuroscientist) and Walter Pitts (logician) [9–11]. A NN is given by "neurons" assuming Boolean values and having entering and exiting edges. Input neurons take input signals from outside in their entering edges, while output neurons send externally along their exiting edges response signals as output. Internal neurons are connected with other neurons by internal edges (synapses). At each step, any neuron receives values along its entering edges, applies to them a **transfer function**, called also **activation function**, and sends the result to the connected neurons through its exiting edges (that are entering edges of the receiving neurons). In McCulloch and Pitts's original version, if the sum of entering values is greater or equal to a given threshold, the neuron sends result 1 along its exiting edges, otherwise, it sends result 0. Some entering edges can be inhibitory and if at least one inhibitory edge enters value 1 into a neuron, it sends 0 as its result.

McCulloch and Pitts's model was shown to be computationally equivalent to finite-state machines [11]. It is interesting to remark that the first project of a modern electronic machine (program driven), elaborated in 1945 by John von Neumann, and called EDVAC (Electronic Digital Variable Automatic Calculator) was based on circuits directly related to McCullough and Pitts neural networks.

The initial NN model, conceived within the classical paradigm of computing machines, evolved in time through different variants, usually called ANN (Artificial Neural Networks), for their perspective, related to artificial intelligence, in solving problems typical of intelligent behaviours: classification, recognition, and constraint satisfaction.

Further versions of neural networks extended the values of neurons from Boolean to real values, and edges were provided by parameters (negative parameters play a role in inhibition). Transfer functions became real functions of many arguments and of different types (sigmoid-like functions are very usual). In these ANN, transfer functions apply to the values arriving along entering edges, after weighting them by the parameters of the edges, and the result is sent along the exiting edges [12].

In more than fifty years of research in ANN, many structural and algorithmic principles were elaborated [13–22] that have provided a surprising efficiency and a wide gamma of applications. Texts [12, 23] are recommended for first-level and second-level introductions on ANN, respectively. Outlines of the history of AI can be found in [24]).

More powerful ANNs exhibit intelligent competencies, including abilities in performing: complex forms of: (1) association, (2) accommodation, (3) assimilation, (4) integration, (5) selection, (6) memorisation, (7) abstraction.

Association is the ability underlying any activity of classification or recognition. Accommodation is the ability to adapt an acquired knowledge for coping with a new situation, while assimilation expresses the ability to reconstruct an internal model of an external reality for elaborating an adequate reaction to it. Integration is the ability to combine abilities already developed. Selection refers to the search for values that satisfy some given constraints. Memorisation is the ability to retain traces of all the previous abilities in order to access them, as directly available when the environment requires them. Abstraction is the ability to eliminate details while keeping the main features of a given pattern according to the needs of a situation of interaction.

The following figures want to support the neurological intuition on which ANNs are based, but also the consideration that their nature is similar to many different kinds of biological networks: metabolic networks, gene activation networks, or social interaction networks. All these structures have a common mathematical basis, even if with many possible variants: a graph formed by nodes (of many possible types) and edges (of many possible types) connecting nodes. It is important to remark that in the picture of Fig. 6.5, for a neuron there are many entering edges and many exiting edges, but exiting edges send all the same output, therefore the information elaborated by each neuron has a flow, directed from inputs to outputs, which corresponds to a function of many arguments providing one result (Figs. 6.1, 6.2 and 6.3).

In the network of Fig. 6.6, the substances associated to two circles provide oscillating values, obtained by using null initial values and suitable linear functions as "regulators" (values of two regulators acting on the same reaction are summed) [25]. Moreover, the graph of Fig. 6.6 is a second-order graph, because meta-edges connect nodes with edges. Graphs and multi-graphs of any finite order can be easily defined. It is easy to show that these kinds of graphs can be completely encoded in simple (first-order) graphs (nodes-edges, or objects-arrows) by using suitable labeling of nodes and edges. For example, the multi-edge of Fig. 6.4 can be seen as a simple

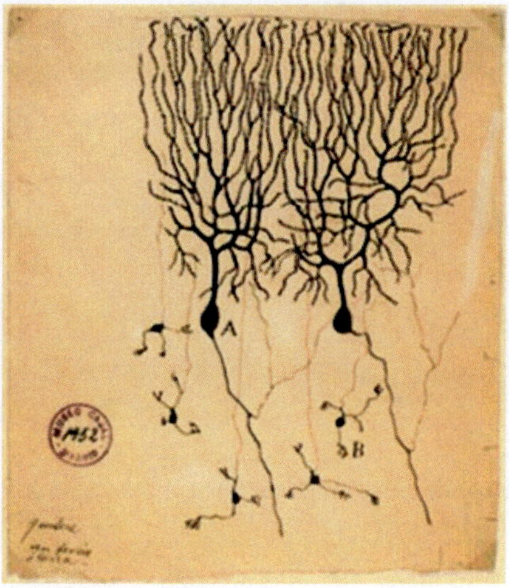

Fig. 6.1 Cajal's model of neurons

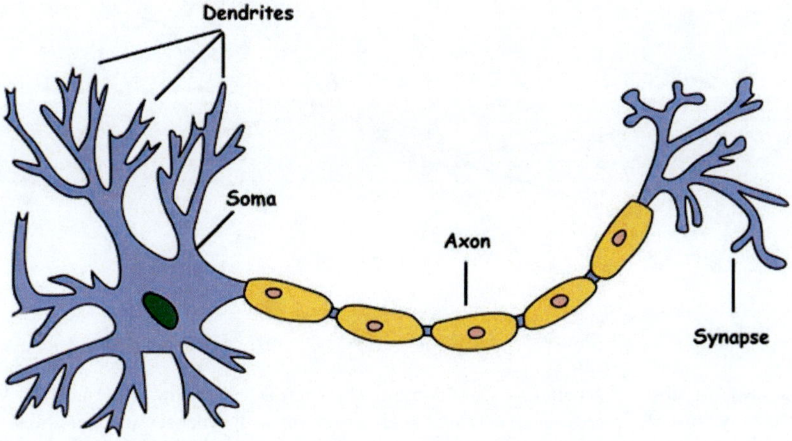

Fig. 6.2 The schema of an internal neuron. The flow of elaboration goes from left (dendrites) to right (synapses), along the axon that sent its signal to axon terminals (Wikipedia)

graph of eight nodes where the central node is of a different type with respect to the others.

Metabolic networks can compute all functions computed by Turing machines [27, 28]. But, it is interesting that they can represent an algorithm by spreading the whole structure, by means of suitable initial values of substances and by the functional forms of their regulators. Regulators are functions that establish the quantity of matter passing through edges. They correspond to the transfer functions of ANN. MP grammars, or MP systems, are a class of metabolic networks with great expressive power in describing many kinds of biological dynamic [29–34], which can be

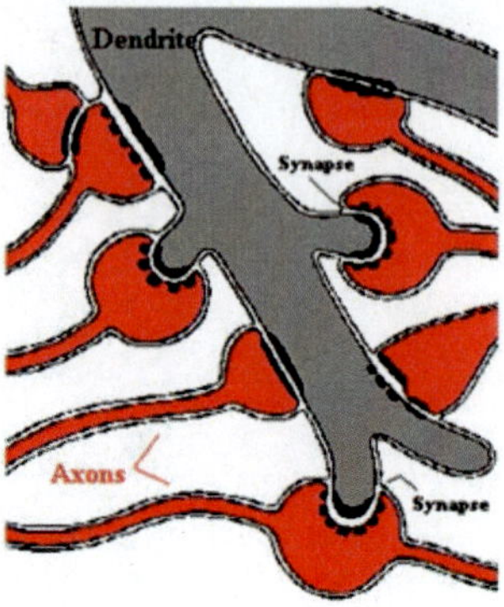

Fig. 6.3 Dendrite-Synapse connections (Wikipedia)

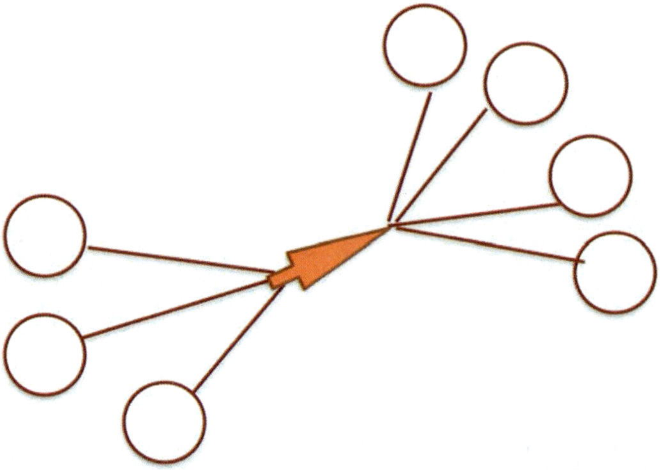

Fig. 6.4 Neuron abstract schema as a multi-edge. The central base arrow is the neuron. White circles are “synapses”. The edges connecting the arrow (axon) with synapses are “dendrites”

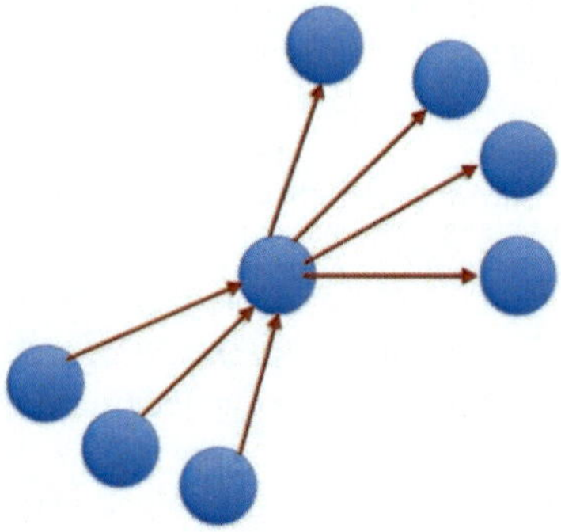

Fig. 6.5 Schema neurons-synapses. This representation is simpler than that of Fig. 6.4, but conceptually equivalent to it: circles are neurons and synapses are edges

Fig. 6.6 A graph modelling an oscillating metabolism. Substances are triangles (input and output substances) and circles. Edges are reactions and wavy edges are "meta-edges" expressing regulators. Any regulator is a function of its source nodes giving, in each state, the quantity of substance transformed by its target reaction

Fig. 6.7 A metabolic network. Substances are triangles and rectangles, reactions are ellipses, and regulators are stars

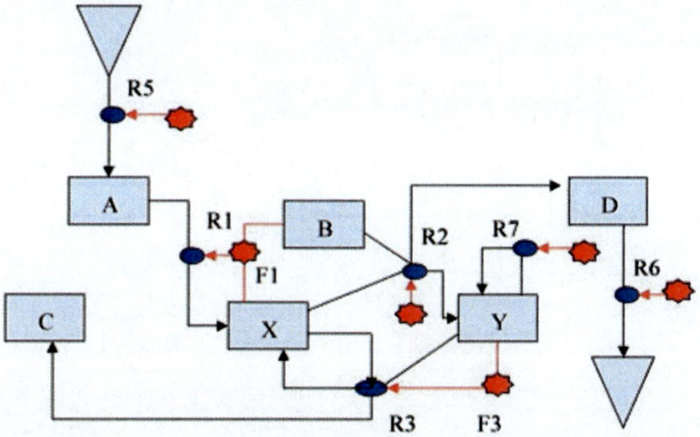

translated into suitable ANNs and *vice versa*. Moreover, any gene activation network can be expressed by a suitable metabolic network [35] (the metabolic network of Fig. 6.7 was derived by a phenomenon of gene activation [36]). This means that the three main types of biological networks show a common deep nature in their mathematical structures. essentially represented by systems of equations, as it is shown in the simple case of the ANN in Fig. 6.9 (Fig. 6.8).

In the general schema of an ANN, nodes are divided into three groups: input neurons, internal neurons and output neurons. The first group receives input external signals, while the last group sends their values as external signals.

Output neurons are responsible for producing the desired outcome of the model. For example, if the ANN is developed for a classification process required to assign one among two classes, then the output neurons codify inputs in one of two classes.

Internal neurons are the key elements of an ANN. They are responsible for processing the input signal into the output values by means of intermediate evaluations.

In general, an ANN implements a mathematical function obtained by combining the transfer functions of its neurons, according to the parameters associated with their edges.

A radical change in ANN research was due to the **Machine Learning** approach (ML), which introduced a sort of inversion in the link between computation and function. In classical models of computation, devices' computing functions start from

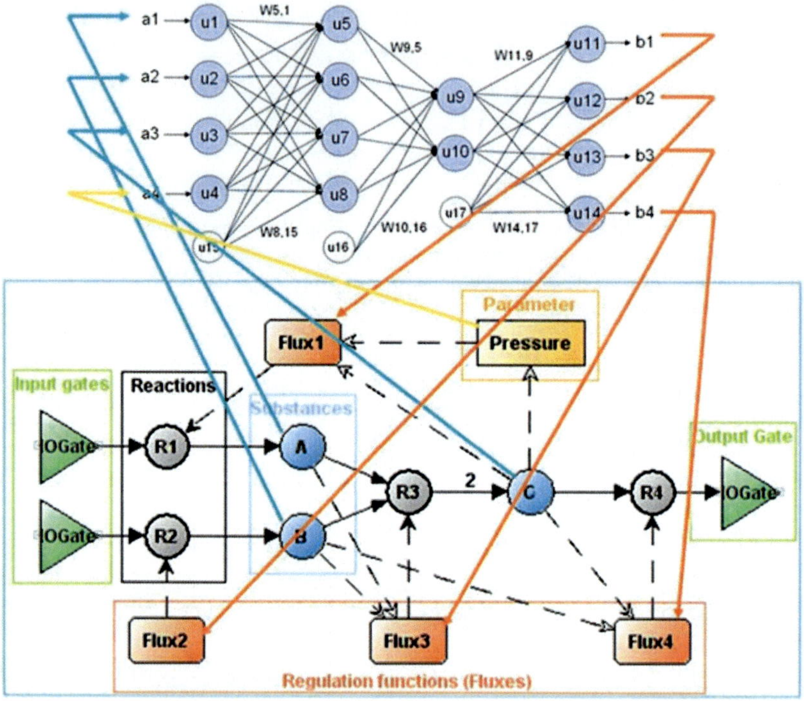

Fig. 6.8 A metabolic network, where metabolic fluxes are controlled by ANNs (from [26])

data by providing, at the end of computation, some associated results. Differently, a machine learning system does not compute the results associated with some data, but infers a function that fits, within some approximation threshold, a given set of data-results pairs. In this way, the function is not the knowledge internal to the machine but it is the result of its computation that is derived according to a suitable strategy from a data set of examples. Of course, this approach naturally introduces viewpoints based on approximation and probability.

The term "Machine Learning" was coined in 1952 by Arthur Samuel, a computer scientist at IBM and a pioneer in AI and computer gaming [37] (see [38], for a classical textbook on ML). The inversion of the Turing computational paradigm is intertwined with statistical and probabilistic aspects. In a sense, this it is rooted in Shannon's approach to information, based on the intrinsic relationship between digital and probabilistic information [39, 40].

The training phase of a learning ANN aims at minimising the error that the ANN produces by its output neurons, with respect to a function that the network is required to compute for acquiring a given competence. Therefore, an error function is defined for expressing the gap between expected and computed functions. Such an error is globally minimised by successive training iterations in a conceptual learning procedure (elaboration of rules and strategies from examples). The main algorithm for realizing such a procedure is **backward propagation**, or simply **backpropagation** [23, 41]. The basic idea of such a procedure is that of propagating errors on output neurons to neurons preceding them in the ANN structure. In this way, the parameters of the mathematical functions implemented by the neurons are iteratively optimised. Backward propagation produces high-performance ANN but comes with a very high computational cost, therefore many variants have been proposed, which usually gain more efficiency by using probabilistic strategies.

Backpropagation, is an old idea, based on the gradient of functions and related to least square approximation, going back to Augustine Cauchy, Jacques Hadamard, and Richard Courant. It is called "method of steepest descent (for non-linear) minimization problems)" in a paper of 1944 by Haskell Curry (the inventor of Combinatory Logic) cited in [42], where a stepwise version of the method is outlined. Paul Werbos proposed in 1974, in his dissertation, the process of training artificial neural networks through backpropagation of errors. The idea was applied, in 1986 in a paper [19], which had an enormous impact in the community of ANNs. It was the algorithmic ingredient that catalyzed a key passage from "computing machines" to "learning machines". If we consider that the least square approximation goes back to Adien-Marie Legenddre and Friedrich Gauss, we can realize as backpropagation is rooted in classical methods of mathematical tradition.

The backpropagation algorithm can be applied to any systems of equations involving derivable functions, therefore its ability in learning parameters can be applied to ANN but also to a wide class of systems, for example, metabolic or gene activation networks [43].

The machine learning approach has promoted a richer relationship between computation theory and the wide area of continuous mathematics related to calculus, probability, approximation, and optimization. Discrete aspects of data representation and programs interact with typical concepts of continuous mathematics. Namely, backpropagation is based on the computation of gradients, and Jabobian matrices, which are basic notions of differential calculus, and the automatic computation of gradients was an essential ingredient of backpropagation [44]. This change of perspective can be expressed by the passage from the universality of Turing machines (Turing-Church thesis) to the universality theorem, telling that any continuous functions from real vectors to real vectors can be approximated by some ANN with internal layers (see [41] for a very intuitive proof), The first universality result which is a consequence of backpropagation. Functions, which are the abstract mathematical notion behind computations, are learned from data by ANN, and any complex activity, going from recognition to translation, can be expressed by a function approximating some suitable continuous many-valued real function.

Recently, machine learning techniques have improved by using **deep neural networks**, in which neurons are structured into layers, and the number of such layers is relatively high. Deep neural networks, often provided by probabilistic mechanisms, improve in accuracy and efficiency ANN capability of approximating continuous functions [23, 45]. This approach was possible through the introduction of hardware, in particular, General-Purpose Graphic Processing Units (GP-GPU) with computation capabilities that were previously unavailable.

Deep ANNs have revolutionised the field of machine learning. However, their best contribution has probably been given by their integration with search algorithms. In fact, their application to learning random distributions, integrated into the Monte-Carlo Tree Search approach, constitutes the core strategy of AlphaFold [46], the system which solved the 50-year-old problem of protein folding. It can predict the spatial form of a protein from the sequence of its amino acids. The related algorithm was developed by DeepMind, soon after its AlphaGo, which in 2016 beat world Go champion Lee Sedol at the game. AlphaGo represents the state space as a tree and explores it partially. Given a state, the next state is chosen according to a heuristic which aims at predicting, in a very efficient way, the gain or the loss in choosing such a move for future states. The heuristic is driven by a simulation which results in a suitable probability distribution.

Neurology can suggest further developments of actual ANN, which, at the same time, can clarify neurological mechanisms in the path toward consciousness. For example, Eric Kandel [47] (Nobel laureate) developed in the last century a theory, supported by many experiments, on a primordial organism, called Aplysia, by showing, in a basic circuit of only four neurons, a phenomenon of general relevance, found in other animals. Namely, the repeated activation of a neuronal circuit, under some specific stimuli, provides the development of a new synapse that in a certain sense encodes the activated circuit promoting its rising. Therefore it results in a memory of acquired competence. The process involves a complex interaction of many subtle biological aspects: the activation of some genes in a neuron, the synthesis of a protein promoting the growth of the synapse, through its peripheral localisation, and as a consequence of the specific nature of this protein, being a prion, the assumption of a stable form that propagates to all proteins of the same kind, which enforces a perpetual form associated to the event causing the circuit activation.

ANNs can be seen as systems of equations. In fact, let us associate a variable X to each neuron, and let us denote by $X(i)$ the value that the neuron assumes in correspondence to inputs i. If f is the transfer function of the neuron, and $w_1, w_2, \ldots, w_n$ the weights of its entering wedges, which are also exiting edges of neurons $X_1, X_2, \ldots, X_n$, then the value $X(i)$ is given by:

$$X(i) = f(w_1X_1(i), w_2X_2(i), \ldots, w_nX_n(i))$$

while input neurons directly take value from outside and output neurons send their value externally. By putting together the equations associated with all neurons, a system of equations is generated that completely expresses the behaviour of the ANN. The number of levels of internal neurons gives the degree of connectivity of the ANN. Assuming that transfer functions are the sum of weighted inputs, the following system of equations corresponds to the ANN of Fig. 6.9 with inputs a, b.

$U(a, b) = w_1X(a, b) + w_2Y(a, b)$

$V(a, b) = w_3X(a, b) + w_4Y(a, b)$

$W(a, b) = w_1X(a, b) + w_6Y(a, b)$

$Z(a, b) = w_7U(a, b) + w_8V(a, b) + w_9W(a, b)$

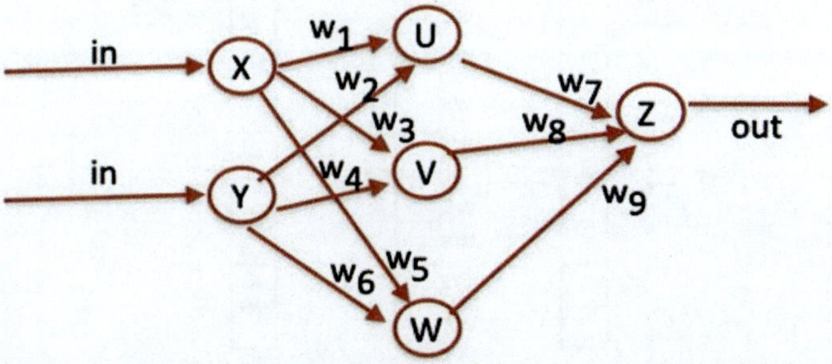

Fig. 6.9 ANN with 6 neurons 12 edges 2 input neurons and 1 output neuron. The system of equations associated with it is given above. Input neurons correspond to variables that occur only in the left members of equations, while output neurons correspond to variables that occur only in the right members

In this representation, an ANN is completely described by a text, or equivalently, by a piece of code of a programming language based on equations (and even training procedures producing the ANN can be suitably encoded by a text).

When the mathematical form of the transfer functions is defined, all the information of the ANN is encoded by the weights occurring in the system of equations. This means that if the ANN is trained to learn a specific function F, relating input neurons with output neurons, this function is entirely given by the values of the weights of the ANN, which memorise the whole competence acquired by the ANN in the learning process.

In the equation representation, it appears, even more clearly, that if a transfer function expresses an elementary mechanism of stimuli-response, the whole ANN is a suitable composition of a great number of simple stimuli-response mechanisms that in the ANN composition become able to develop very complex functionalities, and the final weights record the history of their determination along the whole process of successive adjustments to the data received in the learning process.

Figure 6.10 shows a multi-layer representation of ANN as a network of vectors (layers). It can be applied to represent also metabolic networks, and by means of it, any ANN can be easily transformed into a metabolic network and *vice versa*, This means that the two kinds of structures are equivalent. Figure 6.11 is a pictorial representation of the information flow in a multi-layer representation.

In order to better understand the logic of the learning mechanism via backpropagation, it is useful to consider the metabolic network given in Fig. 6.6. Let us associate the variables X, Y (left, right) to the substances represented by the circles, and assume that regulators are linear functions. We obtain the following system of equations, for $i = 1, 2, \dots 1000$ and $\tau = 2\pi 10^{-3}$ $(X(0) = 1, Y(0) = 0)$:

$$\begin{aligned} X(i\tau) - X((i-1)\tau) &= w_1 + w_2 X(i\tau) - w_3 Y(i\tau) \\ Y(i\tau) - Y((i-1)\tau) &= w_4 X(i\tau) + w_5 Y(i\tau) - w_6. \end{aligned} \quad (6.1)$$

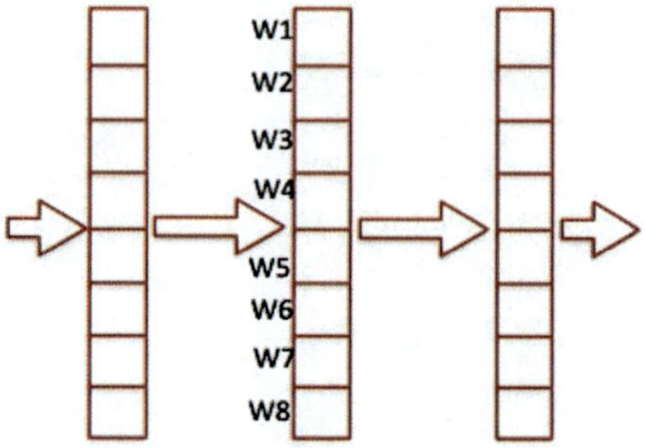

Fig. 6.10 A multi-layer representation of an ANN with one input vector, one internal and one output vector (all of 8 components). Any element of a vector is labelled by a variable that depends (via some basic transfer function) on the components of the previous (left) vector, and the value of the variable, at any step, gives the component of the vector (Playground free image)

Fig. 6.11 A visualization of the flow of information in ANN multi-layer representation

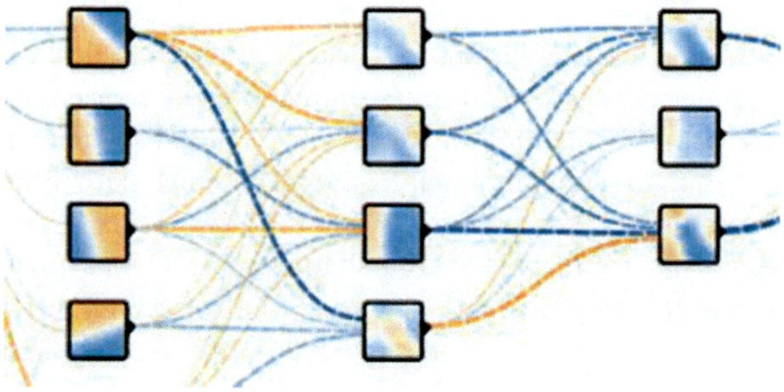

In [25], in the context of a class of metabolic networks, called MP grammars (also metabolic grammars or MP-systems [48]) the values of weights are found such that X and Y approximate the cosine and sine functions, respectively, by means a regression algorithm based on Least Square Evaluation (see [25, 49, 50] Chap. 3, [27, 39, 51] for the use of MP grammars in the approximation of real functions).

The entire time series of the pairs $(i\tau, (X(i\tau), Y(i\tau))$, for $i = 1, 1000$ is considered in such a way that one thousand instances of equations (6.1) are generated, where the right members are equated to cosine and sine values (for $\tau i,\ i = 1, 1000$). The best values of the six weights, according to the least square evaluation, are determined that minimize errors between right members of equations (6.1) and cosine-sine corresponding values.

The performance of this method is very good, even if compared with available ANN learning the same functions, reaching an approximation order of 10^{-14}.

In the backpropagation methods, an analogous determination is obtained, but not at once, rather, in an incremental way of iterative adjustments (driven by the gradient of an error function with respect to the weights). Whence, the right weights are the result of an incremental accommodation to all the pairs arguments-result given during the training of the network. However, the sequential nature of the training, developed along a sequence of single steps, entails that the whole process takes into account, in a cumulative way, the data provided during all the steps. In the case of backpropagation, We could exemplify that instances of equations like (6.1) are solved, with respect to weights, by adding at each step a new instance, where new weights are determined by evaluating the error that previous instances produce with the old values of weights.

The example of the sine-cosine metabolic network also shows, in a very simple setting, other two important aspects common to the "inverse methods" in function determination: (1) an unknown function is determined by combining basic functions (linear, in this case); (2) the six constants determined by the method result to be a sort of compression of all information of the thousand data given for obtaining their determination.

Deep ANNs show us the importance of complex levels of connectivity in order to reach good approximation in the approximation of real functions, related to specific functionalities. This aspect is an a posteriori evidence of the complexity of metabolic transformations that we found in nature. In fact, the more a metabolic network is complex, the more it is efficient, stable, and adaptive. The most part of the complexity of the stepwise regression algorithm given in [25] is the choice of the best "regressors", that is, the basic functions, to put as regulators. This task was realized for metabolic networks [25, 50] by a kind of F-test driving the algorithm to the best choice of regulators. However, as deep ANN clearly indicate, it is better to increase the network, by leaving simple functions acting at the single neurons, rather than having more complex basic functions in smaller networks.

6.4 Artificial Versus Human Intelligence

Recent successes of artificial intelligence (Machine Learning, Neural Networks, Evolutionary and Distributed Computation, Computer Vision, Robotics, Expert Systems, Speech Processing, and Natural Language Processing) in many fields of application raise many kinds of debates about limits, perspectives, and dangers that can arise in the interaction between human and artificial intelligence, in the next future, when cognitive and social involvements of these interactions will be more complex and invasive.

In particular, a new class of systems, called **chatbots** (chat + robots) conquered the attention of mass media for recent decisions of stopping and controlling their diffusion and use in public contexts. These systems can generate texts, from suitable lists of topics of interest (and lines of expansion), by a competence extracted from large moles of data, and exhibit composition/conversation competencies that in some cases seem indistinguishable from those of texts produced by well-educated people with an average competence and knowledge in the specific topics under consideration.

ChatGPT, a very popular chatbot, when requested of defining itself, answers in this way:

GPT (short for "Generative Pre-training Transformer") is a type of language model developed by OpenAI. It is a neural network trained to generate human-like text by predicting the next word in a sequence given a large dataset of examples. Chat GPT refers to the use of this model in chat applications, where it can be used to generate responses to user messages in real-time. It is designed to be able to carry on a conversation by understanding the context of the conversation and generating appropriate responses.

The last versions of GPT (GPT-4, 2022) are based on neural networks with a size comparable to the human brain (trillions of neurons and around 100 trillion parameters). The emergence of the ability shown by these systems was a real surprise for many people, and up to now many aspects are not fully understood, hence the comparison with the size of the human brain is very interesting, but more advanced analyses need to be developed in this regard: a human brain has a massively parallel configuration of 10^{12} neurons and 10^{15} synapses in less than two litres, with a weight of 1 kg and operate at low power of less than 20 W. However biological neurons have structure extremely complex with respect to artificial neurons.

A chatbot can write essays, solve coding problems, translate texts, and even discuss topics of interest. For it (her/him) producing coding in Python is very easy, and even if the first answer can be incorrect, reformulating the question or reporting the error messages of some generated coding, in the end, very often, produces correct programming texts useful to solve the problem posed to it. In fact, in a continuous conversation with it, previous interactions are remembered by it, hence from a sequence of follow-up questions a final answer can be obtained that gives a working piece of code.

However, in some cases, errors (due, for example, to homonymy) may generate products that can be inappropriate or offend the privacy of mentioned persons. For this reason, some national authorities took actions aimed at contrasting the free diffusion of some chatbots. These episodes suggest general speculations about the essence of comprehension and intelligence.

In a sense, chatbots are generalizations of an old idea introduced by Alan Turing, about a behavioural approach to the definition of comprehension. Turing's Test was realized by a program chatting with a human, through some interface by means of which two agents develop a dialogue. One of them is human and he is requested to manifest, along the conversation, when he got the impression of speaking with an artificial system misunderstanding the terms of their dialogue. In a sense, the average length of conversations where the human does not receive any misunderstanding impression can be considered a measure of the "conversational ability" of the artificial system. The natural question in this regard is that of establishing if a system able to develop a safe dialogue can be considered able to comprehend, and in this case, in which terms his/her comprehension could reasonably be defined.

Presently, chatbots are not equipped with any active memory of the whole conversational processes that they are able to develop. However, this does not mean that future versions of these systems cannot acquire such kind of ability. The main criticism of the cognitive competence of chatbots is the consideration that they realize only a syntactic play, learned through a training process based on a huge quantity of data, but with no real semantic acquisition. Even if this is true, it is also true that in complex systems languages emerge by syntactic games that develop and acquire meanings in the interaction process among communicating agents. In formal languages, semantics is defined independently from syntax, and interpretation rules assign meaning to syntactic forms (well-formed formulae) defined independently from semantics. In natural languages, the interaction between syntax and semantics is constructed during the emergence and use of languages, and usually, semantics follows syntax. Child experience in developing linguistic activities gives many examples of mechanisms acquired without any full understanding of related meanings. According to this perspective, a chatbot who associates his/her interaction with the development of a neural system tracing functions and correlations of his/her activities could be an agent where the dialogue play generates primordial forms of comprehension. In any case, memorization is a necessary acquisition of any comprehension. Individual identity and consciousness are strongly based on memory, as recent acquisitions of neurological research definitively confirm.

In the previously outlined basic model of "comprehension machine", another essential ingredient has to be introduced which is "the fuel of the motor". In fact, no comprehension can be formed without a will to push toward it. In the case of many learning processes, what moves the attention of the learning agent is some emotional gratification based on affective interactions. This means that a chatbot learning the meanings of what he/she says, through a suitable neural network (making connections and selecting functional parameters), needs to be able to develop an emotional system supporting the coupling between syntax and semantics (with primitive forms of memory and consciousness). Pain, joy, hope, and love are necessary for any real understanding. In conclusion, a body with all its sensory and perceptive activities is necessary for full comprehension, in the absence of such a corporal part, any AI agent can provide only an *artificial, abstract, or assisted* comprehension. Namely, it is well known that, in child educational activity, affective deficits preclude even logical and mathematical abilities because the attention requested for dominating abstract ideas needs emotional support to attract and fix attention on the topics of a learning activity.

It is too early to say if and when some future "very intelligent" possible scenarios of AI could be, even partially, realized. In principle, it seems only a question of increasing and integrating competencies that singularly are, in some forms, already available in the actual information technology. However, realizations could be blocked by quantitative limits in computational abilities, or by problems arising with the scale-up in complexity as far as many intelligent activities are integrated into autonomous individual systems.

A remarkable aspect of the relationship between humans and artificial intelligence is the impact that artificial intelligence systems have in producing new forms of human intelligence, and, possibly, in helping to reach more accurate models of human intelligence.

The educational processes with which human knowledge is transmitted and human minds are formed, by acquiring the individual and collective consciousness of a culture, can dramatically change by a massive presence of artificial systems assisting humans in many kinds of tasks.

In particular, the classical paradigm of mathematical proof, where passages are completely dominated by human intelligence, changes radically in situations where solutions are found by AI agents (Artificial Intelligence agents), trained by humans, where no complete knowledge is available for the single steps performed by the machine for reaching a solution. Along this way, the human approach to problems surely is influenced and even the meaning of "comprehension" assumes new characters, when human knowledge includes pieces that are not directly accessible.

Let us report some fragments of recent conversations we had with ChatGPT. From the OpenAI site (What is ChatGPT?) we report the following statement. "Subject to the Content Policy and Terms, you own the output you create with ChatGPT, including the right to reprint, sell, and merchandise "regardless of whether the output was generated through a free or paid plan."

A conversation with ChatGPT

User
what is ChatGPT?
ChatGPT
Hello! I am ChatGPT, a large language model developed by OpenAI. I am based on the GPT (Generative Pre-trained Transformer) architecture, specifically the GPT-3.5 version. I have been trained on a massive amount of text data and can understand natural language inputs, generate responses, and perform various language-related tasks. Essentially, I am an AI-powered chatbot designed to communicate with humans in a way that feels natural and conversational.

User
What is a Transformer architecture?
ChatGPT
Transformer is a neural network architecture that was introduced in a paper titled "Attention Is All You Need" by Vaswani et al. in 2017. The Transformer architecture is based solely on the attention mechanism, which allows the network to selectively focus on different parts of the input sequence during processing.

Unlike previous neural network architectures, such as recurrent neural networks (RNNs) or convolutional neural networks (CNNs), the Transformer architecture does not rely on sequential processing. Instead, it processes the entire sequence of input tokens at once, enabling it to parallelize training and inference and making it more efficient than traditional RNN or CNN models.

The Transformer architecture has been widely adopted in various natural language processing (NLP) tasks, such as machine translation, language modeling, and text classification. In fact, the GPT-3, one of the largest and most powerful language models to date, is based on the Transformer architecture.

User
What is an "attention mechanism"?
ChatGPT
An attention mechanism is a key component of many neural network architectures, particularly those used for natural language processing tasks. It enables the network to selectively focus on different parts of the input sequence during processing, allowing it to process the most relevant information while ignoring irrelevant or noisy information.

In the context of neural networks, attention can be thought of as a way of assigning weights to different parts of the input sequence based on their relative importance to the task at hand. These weights are learned during training and are used to weight the contribution of each input token to the final output of the network.

One of the most commonly used attention mechanisms is called "self-attention" or "scaled dot-product attention", which is used in the Transformer architecture. In self-attention, each input token is transformed into a query, a key, and a value vector. The query vector is then used to compute a weighted sum of the value vectors, where the weights are determined by the similarity between the query vector and the key vectors. This weighted sum is then used as the output of the self-attention mechanism.

Overall, attention mechanisms have proven to be a powerful tool for modelling complex, sequential data such as natural language text, and have enabled significant improvements in the accuracy of many NLP tasks.

User
What is a token?
ChatGPT
In natural language processing (NLP), a token refers to a sequence of characters that represents a single unit of meaning. Typically, tokens are words, but they can also be phrases, numbers, punctuation marks, or other types of elements that convey meaning in a language.

Tokenization is the process of splitting a text into individual tokens or words. This is an important step in many NLP tasks, as it allows the text to be processed by algorithms that can only operate on structured data. For example, in a machine learning model that predicts the sentiment of a text, the text must first be tokenized into individual words or phrases before it can be fed into the model.

In some cases, tokenization can be a non-trivial task, particularly in languages where words are not separated by spaces (such as Chinese or Japanese) or where compound words are common (such as German). In these cases, specialized algorithms may be needed to accurately tokenize the text.

User
Does this mean that a sentence is transformed into a sequence of tokens corresponding to the words of the sentence?
ChatGPT
Yes, that's correct! When a sentence is tokenized, it is typically transformed into a sequence of tokens, where each token represents a word in the sentence. For example, the sentence "The quick brown fox jumps over the lazy dog" might be tokenized into the following sequence of tokens:

["The", "quick", "brown", "fox", "jumps", "over", "the", "lazy", "dog"]

Note that tokenization can sometimes involve additional processing steps, such as lowercasing all the tokens or removing punctuation marks, depending on the specific task and requirements. However, the basic idea is to break down the input text into smaller, more manageable units that can be processed by machine learning algorithms or other NLP tools.

User

Is each token a vector of features?

ChatGPT

Yes, typically each token is represented as a vector of features in natural language processing (NLP) tasks. These features can be thought of as numerical representations of various aspects of the token, such as its spelling, part of speech, or context.

The specific set of features used to represent a token can vary depending on the task and the specific NLP tool being used. For example, in a sentiment analysis task, the features might include the token's frequency in a corpus of positive or negative reviews, while in a named entity recognition task, the features might include the token's capitalization and the presence of certain keywords.

One common way to represent a token as a feature vector is to use a one-hot encoding, where each feature corresponds to a specific attribute of the token and is represented as a binary value (0 or 1). Another approach is to use word embeddings, which are dense, low-dimensional vectors that capture the semantic relationships between words based on their usage in context.

Regardless of the specific representation used, the goal is to transform the text into a format that can be processed by machine learning algorithms or other NLP tools, allowing us to extract useful information and insights from the text.

User

How long is the vector associated with a word, and are all vectors of the same length?

ChatGPT

The length of the feature vector associated with a word can vary depending on the specific NLP tool being used and the representation being used for the word.

For example, in a one-hot encoding representation, the feature vector for a word is typically a vector of 0's and 1's, with a length equal to the size of the vocabulary being used. Each element of the vector corresponds to a specific word in the vocabulary, and the element corresponding to the word being represented is set to 1, while all other elements are set to 0. So, in this case, the length of the vector would be equal to the size of the vocabulary.

In contrast, in a word embedding representation, the feature vector for a word is typically a dense vector of floating-point numbers, with a fixed length that is chosen by the user or determined by the NLP tool being used. Commonly used word embedding models such as Word2Vec or GloVe typically produce word vectors of length 100, 200, 300, or 400.

Regardless of the specific representation used, it is important that the vectors are of the same length in order to enable the data to be processed by machine learning algorithms or other NLP tools. In order to achieve this, some NLP tools may pad shorter vectors with zeros or truncate longer vectors to a fixed length, depending on the requirements of the specific task.

User

How many languages can GPT speak?

ChatGPT

GPT models are trained on large amounts of text data from a wide variety of sources and are designed to be able to process and generate text in many different languages. However, the extent to which a GPT model can "speak" or generate text in different languages depends on the specific language(s) that the model has been trained on.

The original GPT model was trained on English text data, but subsequent versions of the model, including GPT-2 and GPT-3, have been trained on much larger and more diverse datasets that include text in many different languages. GPT-3, for example, was trained on a dataset of over 570GB of text data in multiple languages, including English, Spanish, French, German, Chinese, and many others.

While GPT models are not explicitly designed to be multilingual, they are able to generate text in multiple languages based on the patterns and relationships they have learned from the training data. However, the quality and fluency of the generated text may vary depending on the specific language and the amount of training data available for that language.

User

When you say that GPT understands a sentence, what do you mean in more precise terms?

ChatGPT

When we say that GPT "understands" a sentence, we mean that it has the ability to process and represent the meaning of that sentence in a way that can be used for downstream natural language processing tasks.

More specifically, when a sentence is input to a GPT model, it is first tokenized into a sequence of word embeddings, each of which represents a specific word in the sentence. The model then processes these embeddings through multiple layers of self-attention and feedforward networks, gradually building up a more abstract representation of the sentence that captures its meaning and context.

At each layer, the model learns to attend to different parts of the input sequence and to integrate information from previous layers to form a higher-level representation. This allows the model to capture both local and global dependencies between words and phrases in the sentence, and to extract patterns and relationships that can be used to represent its meaning.

Once the input sentence has been processed through the GPT model, the output is a final hidden state vector that represents the entire sentence. This vector can be used as input to downstream tasks such as language modelling, question answering, or text classification, where it can be used to generate predictions or make decisions based on the meaning of the input text.

While we often use the term "understands" colloquially to describe this process, it is important to note that GPT and other language models do not truly "understand" language in the way that humans do. Instead, they learn to recognize patterns and relationships in large amounts of text data and to use this knowledge to perform specific NLP tasks.

User

Ok, you say "hidden state vector", please, could you explain better the attribute "hidden"?

ChatGPT

The term "hidden" in "hidden state vector" refers to the fact that this vector represents the internal or hidden state of the GPT model after it has processed the input sentence. The vector is not directly visible or accessible to the user, but it contains valuable information about the meaning and context of the input text.

In the case of GPT models, the hidden state vector is typically a high-dimensional vector with several hundred or thousand dimensions, depending on the specific architecture and hyperparameters of the model. This vector is computed through a series of computations and transformations that take place as the input sentence is processed through the model.

Despite being "hidden", the information contained in the hidden state vector is extremely useful for downstream NLP tasks, such as text classification or language modelling. By using this vector as input to another model or algorithm, we can leverage the contextual information captured by the GPT model to improve the accuracy and performance of our task.

It's worth noting that the term "hidden" can be a bit misleading, as the hidden state vector is not actually a secret or mysterious part of the model. Rather, it is simply a term used to describe the fact that this vector is not directly observable or interpretable in the same way that the input tokens or output predictions are.

User

Can we approximately say that the hidden state vector associated with a statement understanding is constituted by the weights of GPT's ANN, or of a part of it, at the moment that the representation of this understanding is generated?

ChatGPT

It's not quite accurate to say that the hidden state vector associated with a statement understanding is constituted by the weights of GPT's ANN (artificial neural network), or a part of it, at the moment that the representation of this understanding is generated.

The hidden state vector is actually a computational output of the GPT model that represents the internal state of the model after it has processed the input sentence. It is computed by applying a sequence of mathematical transformations to the input sentence, using the model's weights and parameters, and does not directly correspond to the weights of any specific layer or component of the model.

That being said, the weights and parameters of the ANN do play a crucial role in determining the structure and behaviour of the GPT model and are carefully tuned during the training process to optimize the model's performance on a given task. So while the hidden state vector is not directly constituted by the model's weights, it is ultimately derived from the interactions between these weights and the input sentence and can be seen as a reflection of the model's learned knowledge and patterns of representation.

User
Ok, but when you say "a computational output" what do you exactly mean? Where this output can be localized?
ChatGPT
By "computational output", I mean that the hidden state vector is a result of computations performed by the GPT model during the processing of the input sentence. More specifically, the hidden state vector is produced by the final layer of the GPT model, which is typically a fully connected layer that takes as input the output of the preceding layer and produces a high-dimensional vector as output.

In the case of GPT models, the hidden state vector is often associated with the output of the final layer of the model, which is a fully connected layer that takes as input the sequence of token embeddings produced by the preceding layers. The weights and biases of this final layer are learned during the training process and are optimized to produce an output vector that captures the contextual information and meaning of the input sentence.

Once the input sentence has been processed by the GPT model, the hidden state vector is typically stored in memory or passed to downstream algorithms or models for further processing. The location and format of this output will depend on the specific implementation of the GPT model and the application being used. For example, in the case of the original GPT model, the hidden state vector is output by the final layer of the model as a tensor with dimensions (batch size, sequence length, hidden size).

User
In the multi-head approach of the transformer GPT structure, which is the criterion for establishing how many heads to consider in a statement?
ChatGPT
The number of attention heads used in the multi-head approach of the Transformer-based GPT model is typically a hyperparameter that is set before training the model. The choice of the number of heads depends on several factors, including the size of the training data, the complexity of the task being addressed, and the available computational resources.

In general, a larger number of attention heads can help the model to capture more fine-grained patterns and dependencies in the input sequence, at the cost of increased computational complexity and potentially longer training times. However, there are diminishing returns to increasing the number of attention heads, and there may be a point at which further increases provide little benefit.

Empirically, the number of attention heads used in GPT models has varied across different versions of the model, with some models using as few as 12 heads and others using up to 48 heads. The original GPT model used 16 attention heads, while the subsequent GPT-2 and GPT-3 models used up to 48 and 96 heads, respectively.

Ultimately, the choice of the number of attention heads is a trade-off between model complexity, computational efficiency, and task performance, and is often determined through experimentation and hyperparameter tuning.

User
At the end of the representation process of a statement, is a final vector associated

with the sentence?
ChatGPT
Yes, in many natural language processing (NLP) tasks, a final vector is associated with a sentence after the representation process is complete. This vector is often referred to as a sentence embedding or a sentence vector, and it represents a compressed and abstracted version of the original sentence that can be used for downstream tasks like sentiment analysis, text classification, or machine translation.

There are many different methods for generating sentence embeddings, but most of them involve some form of pooling or aggregation of the word embeddings or feature vectors associated with the individual words in the sentence. For example, one common approach is to take the average of the word embeddings or feature vectors, while another approach is to use a more sophisticated pooling technique like max pooling or self-attention.

The length and dimensionality of the final sentence vector can vary depending on the specific method used and the requirements of the task. In some cases, the sentence vector may be a fixed-length vector with a pre-defined dimensionality, while in other cases it may have a variable length or a dimensionality that depends on the length of the input sentence.

Overall, the goal of generating a sentence vector is to provide a compact, high-level representation of the sentence that can be used by machine learning algorithms or other NLP tools to make predictions or extract insights from the text.

User
How is realized the training of GPT?
ChatGPT
The training of GPT, or any other large language model based on the transformer architecture, involves several steps:

Corpus Collection: The first step is to collect a large corpus of text data that the model will be trained on. This can be done using a web crawler to scrape text from the internet, or by using existing datasets such as Wikipedia or news articles.

Preprocessing: The text data is then preprocessed to clean and normalize the text, and to tokenize it into individual words or subwords. The tokens are typically mapped to integer IDs for efficient processing.

Architecture Design: The transformer architecture used by GPT is designed to process sequences of tokens in a parallel and hierarchical manner, using multi-head attention and feedforward layers. The architecture is often pre-trained on large amounts of text data using an unsupervised learning objective, such as predicting the next word in a sequence or filling in a masked token.

Training: During training, the model is fed batches of sequences of tokens, and the weights of the model are updated using backpropagation and stochastic gradient descent. The goal is to minimize a loss function that measures the difference between the model's predictions and the true target values (e.g., the next word in the sequence).

Fine-tuning: Once the model has been pre-trained on a large corpus of text, it can be fine-tuned on a smaller dataset for a specific task, such as sentiment analysis or text classification. During fine-tuning, the weights of the model are updated using supervised learning, with the goal of minimizing a task-specific loss function.

Evaluation: The final step is to evaluate the performance of the model on a held-out test set, using metrics such as accuracy, F1 score, or perplexity.

Overall, the training of GPT involves a combination of unsupervised pre-training on a large corpus of text data, followed by supervised fine-tuning on a smaller dataset for a specific task. This approach has been shown to be effective for a wide range of NLP tasks and has led to significant improvements in state-of-the-art performance on many benchmarks.

End of the conversation

The entire conversation was a source of basic understanding of the main ideas on the structure and of the functions of the chatbot. Namely, after reading many introductory articles about ChatGPT, the self-presentation was surely the most useful font on the topic. We had other conversations, on different scientific arguments. In one of them, the chatbot was requested to prove Nichomachus' theorem (the square of the sum of numbers from 1 to n is equal to the sum of corresponding cubes) and the proof was correctly stated by using mathematical induction. However, when a Turing machine was requested to duplicate a string given in input, the program exhibited by ChatGPT was wrong, even with some suggestions from the user. Other errors were found in other programming tests, and in technical questions of formal language theory, at an elementary level. However, when the discussion was about the meaning of Turing-Church Thesis, the adequacy of the answers of ChatGPT was very impressive, and an analogous impression was reported when the dialogue was about the universality theorem of ANN's in approximating continuous real functions.

Three considerations are in order. First, AI agents are produced by humans and trained by humans. This means that their knowledge is, in a sense, human knowledge of the second level, even when it realizes activities beyond the direct control of humans. In other words, AI agents can be defined as intellectual prostheses of human intelligence.

Secondly, in a sense from the reported conversation, a sort of model of chatbot comprehension emerges. In fact, a vector of very large dimension can be associated, which is an internal representation of the statement used by the machine in further elaborations for dialogue development. This vector derives from the token vectors associated with the components of the statement, by using dictionaries and attention matrices learned during the learning process.

Thirdly, the knowledge of a chatbot is usually internalized during the training phase, where he/she learns to react correctly to some given inputs within a specific class of data. Extending the training, with a larger or updated set of data, determines a greater competence of the agent. This means that knowledge acquisition is not an autonomous process, but remains a human-driven activity, where a crucial element is missing, that is, an intrinsic internal autonomous and free dynamics aimed at continuously acquiring new knowledge: just the kind of activity, that in a collective manner, pushed humans to realize chatbots.

Finally, our brain is a network of neurons, in some way, written inside our genomes, because all functions of living organisms are encoded in genomes, as any kind of information can be expressed by strings, This information passes from its potential state, at the genome level, to its active state of directing actions, or from syntax to semantics, by means of networks: metabolic, genetic, hormonal, immunological, neurological. The last kind of network provides support to the more complex activities of living behaviours.

Maybe it is now the time that artificial intelligence provides us with new forms of comprehension, by using suitable kinds of ANN for helping humans to understand genomes.

A first suggestion could be the following. Evolution is the essence of life, but its internal logic remains in many aspects unknown. What is really difficult to grasp is its ability in combining casualty with finality. It seems unbelievable that its search in the space of possibilities could be so effective and efficient. In some computational experiments on evolutionary systems [52] it emerges that when these systems have learning abilities, then evolution is much more efficient. Therefore, are there learning mechanisms internal to genomes? In other words, is there some sort of **Genome Learning** that drives evolution? Can machine learning help humans in discovering such genome learning? A very basic observation is that genomes memorize a quantity of information that exceeds the needs of life functions within the organisms. Only a very small portion of the human genome encodes proteins or RNA strands with regulatory or informational roles. Therefore, it is reasonable to infer that the remaining part of genomes can be explained in terms of evolutionary dynamics. Probably, the corresponding information drives learning mechanisms able to finalize the evolutionary steps of genomes, by selecting specific paths. If this assumption is correct, in which way this happens?

References

1. Katoch, S., Chauhan, S.S., Kumar, V.: A review on genetic algorithm: past, present, and future. Multimed. Tools Appl. **80**, 8091–8126 (2021)
2. Lecoutre, C.: Constraint Networks: Targeting Simplicity for Techniques and Algorithms. Wiley, New York (2013)
3. Hell, P.: From graph colouring to constraint satisfaction: there and back again. Algorithms Comb. **26**, 407 (2006)
4. Duan, H., Qiao, P.: Pigeon-inspired optimization: a new swarm intelligence optimizer for air robot path planning. Int. J. Intell. Comput. Cybern. **7**(1), 24–37 (2014)
5. Eberhart, R., Kennedy, J.: A new optimizer using particle swarm theory. In: MHS'95. Proceedings of the Sixth International Symposium on Micro Machine and Human Science. IEEE (1995)
6. Kennedy, J., Eberhart, R.: Particle swarm optimization. In: Proceedings of ICNN'95-International Conference on Neural Networks, vol. 4. IEEE (1995)
7. Glover, F., et al.: Tabu search principles. Tabu Search 125–151 (19970
8. Kirkpatrick, S.C., Gelatt, D., Jr, Vecchi, M.P.: Optimization by simulated annealing. Science **220**(4598), 671–680 (1983)
9. McCulloch, W., Pitts, W.: A logical calculus of ideas immanent in nervous activity. Bull. Math. Biophys. **5**(4), 115–133 (1943)
10. Kleene, S.C.: Representation of events in nerve nets and finite automata. Ann. Math. Stud. **34**, 3–41 (1956). Princeton University Press

11. Minsky, M.L.: Computation: Finite and Infinite Machines. Prentice-Hall Inc., Englewood Cliffs, N. J. (1967)
12. Nielsen M.A., Chuang I.L.: Quantum Computation and Quantum Information. Cambridge University Press (2000)
13. Hebb, D.: The Organization of Behavior. Wiley, New York (1949)
14. Rosenblatt, F.: The Perceptron-a perceiving and recognizing automaton. Cornell Aeronaut. Lab., Report 85-460-1 (1957)
15. Rosenblatt, F.: The perceptron: a probabilistic model for information storage and organization in the brain. Psychol. Rev. **65**(6), 386–408 (1958)
16. Minsky, M., Papert, S.: Perceptrons: An Introduction to Computational Geometry. MIT Press. (1969)
17. Hopfield, J.J.: Neural networks and physical systems with emergent collective computational abilities. Proc. Natl. Acad. Sci. **79**(8), 2554–2558 (1982)
18. Hopfield, J.J., Tank, D.W.: Neural computation of decisions in optimization problems. Biol. Cybern. **52**(3), 141–6 (1985)
19. Rumelhart, D.E., Hinton, G.E., Williams, R.J.: Learning representation by backpropagation errors. Nature **323**, 523–536 (1986)
20. Werbos, P.: Backpropagation through time: what it does and how to do it. Proc. IEEE **78**(10), 1550–1560 (1990)
21. LeCun, Y., Bottou, L., Bengio, Y., Haffner, P.: Gradient-based learning applied to document recognition. Proc. IEEE **86**(11), 2278–2324 (1998)
22. Bahdanau, D., Bengio, Y. Neural machine translation by jointly learning to align and translate (2016). arXiv:1409.0473v7 [cs.CL]
23. Goodfellow, I., Bengio, Y. Courville A.: Deep Learning. MIT Press (2016)
24. Russell, S.J., Norvig, P.: Artificial Intelligence A Modern Approach, 3rd edn. Pearson Education Limited (2016)
25. Manca, V.: Infobiotics: Information in Biotic Systems. Springer, Berlin (2013)
26. Castellini, A., Manca, V.: Learning Regulation Functions of Metabolic Systems by Artificial Neural Networks. GECCO'09, Montréal, Québec, Canada. ACM 978-1-60558-325 (2009)
27. Manca, V.: Metabolic computing. J. Membr. Comput. **1**, 223 (2019)
28. Lombardo, R., Manca, V.: Arithmetical metabolic p systems. In: Foundations on Natural , Artificial Computation, vol. 6686/2011, Springer, Berlin (2011)
29. Castellini, A.: Algorithms and Software for biological MP Modeling by Statistical and Optimization Techniques. Dipartimento di Informatica University di Verona. Ph.D. Thesis (2010)
30. Pagliarini, R.: Modeling and Reverse-Engineering Biological Phenomena by means of Metabolic P Systems. Ph.D. Thesis, Dipartimento di Informatica University di Verona (2011)
31. Marchetti, L.: MP Representations of Biological Structures and Dynamics. Ph.D. Thesis, Dipartimento di Informatica University di Verona (2012)
32. Lombardo, R.: UnconventionalComputations and Genome Representations, Ph.D. Thesis, Dipartimento di Informatica University di Verona (2013)
33. Castellini, A., Manca, V.: Metaplab: a computational framework for metabolic p systems. In: Membrane Computing, WMC9 2008, LNCS 5391, pp. 157–168. Springer, Berlin (2009)
34. Castellini, A., Paltrinieri, D., Manca, V.: MP-GeneticSynth: Inferring biological network regulations from time series. Bioinformatics **31**, 785–787 (2015)
35. Marchetti, L., Manca, V.: A methodology based on MP theory for gene expression analysis. In: Membrane Computing, CMC 2011, vol. 7184, pp. 300–313. Springer, Berlin (2012)
36. Bollig-Fischer, A., Marchetti, L., Mitrea, C., Wu, J., Kruger, A., Manca, V., Draghici, S.: Modeling time-dependent transcription effects of HER2 oncogene and discovery of a role for E2F2 in breast cancer cell-matrix adhesion. Bioinformatics **30**(21), 3036–3043 (2014)
37. Samuel, A.: Some studies in machine learning using the game of checkers. IBM J. Res. Dev. **3**(3), 210–229 (1959)
38. Mitchell, T.: Machine Learning. McGraw Hill (1997)
39. Manca, V.: Grammars for discrete dynamics. In: Holzinger, A. (ed.) Machine Learning for Health Informatics. Lecture Notes in Artificial Intelligence, LNAI 9605, pp. 37–58. Springer, Berlin (2016)

40. Manca, V.: A brief philosophical note on information. In: Holzinger, A., et al. (eds.) Integrative Machine Learning, LNAI 10344, pp. 146–149 (2017)
41. Nielsen, M.: Neural Networks and Deep Learning (2013). http://neuralnetworksanddeeplearning.com/
42. Kelley, H.J.: Gradient theory of optimal flight paths. ARS J. **30**(10), 947–954 (1960)
43. Castellini, A., Manca, V., Suzuki, Y.: Metabolic P system flux regulation by artificial neural networks. In: Membrane Computing, WMC 2009, vol. LNCS 5957, pp. 196-209. Springer, Berlin (2010)
44. Linnainmaa, S.: The Representation of the Cumulative Rounding Error of An Algorithm as a Taylor Expansion of the Local Rounding Errors (Masters) (in Finnish). University of Helsinki (1970)
45. Hornik, K., Stinchcombe, M., White, H.: Multilayer feedforward networks are universal approximators. Neural Netw. **2**(5), 359–366 (1989)
46. Jumper, J., Evans, R., Pritzel, A., et al.: Highly accurate protein structure prediction with AlphaFold. Nature **596**, 583–589 (2021)
47. Kandel, E.R.: In Search of Memory. The Emergence of a New Science of Mind. W. W. Norton & Company, Inc. (2006)
48. Manca, V.: Metabolic P systems. Scholarpedia **5**(3), 9273–9273 (2010)
49. Manca, V., Marchetti, L.: Metabolic approximation of real periodical functions. J. Log. Algebr. Program. **79**, 363–373 (2010)
50. Manca, V., Marchetti, L.: Solving dynamical inverse problems by means of metabolic p systems. BioSystems **109**, 78–86 (2012)
51. Marchetti, L. Manca, V.: MpTheory Java library: a multi-platform Java library for systems biology based on the Metabolic P theory. Bioinformatics **15**; **31**(8), 1328–1330 (2015)
52. Kelly, K.: Out of Control, the New Biology of Machines, Social Systems and the Economic World. Copyright 1994 by Kevin Kelly (2008)

Chapter 7
Introduction to Python

7.1 The Python Language

Python 3 is used as a programming language for this laboratory. The choice of Python is due to three main factors. Firstly, the Python syntax is very intuitive and similar to a spoken natural language, making it among the easiest to learn programming languages. Secondly, it is portable and freely available, thus a Python script can be run on any operating system for which a Python interpreter has been released. Thirdly, as a result of the previously listed factors, Python is one of the most currently used programming languages, especially in machine learning and bioinformatics. In fact, several frameworks and computational instruments are developed on top of the python environment. For all of these reasons, Python results in a very powerful instrument for didactic scopes. If fact, it allows the reader to focus attention on the algorithmic aspect regarding the solution to a computational problem, by ignoring technical details that are specific to a given programming language, such as the management of the memory.

Python has been designed to be a scripting language, thus a Python script is interpreted by an interpreter who reads the script file and executes the instructions line by line. This approach affects the way in which the programmer can write a Python program. The resultant philosophy is that *everything that needs a declaration and that has not been defined in a previous line can not be used in the current line of code*. Typical examples are Python functions, they can not be called by an instruction that precedes the definition of the given called function (by following the reading order of the interpreter). This is one of the main differences between an interpreted and a compiled programming language. Python has been influenced by other programming languages, such as Modula-3, C, C++, Perl, Java and Unix shell programming language. The result is a powerful programming language which presents a very intuitive syntax that hides the technical aspects of more sophisticated languages. The mixture results in a multi-paradigm language which supports procedural programming (which is mainly based on functions and procedures), object-oriented programming (by allowing basic notions regarding classes, such as inheritance, operator over-

V. Manca and V. Bonnici, *Infogenomics*, Emergence, Complexity and Computation 48,
https://doi.org/10.1007/978-3-031-44501-9_7

loading and duck typing) and some aspects of the functional programming (such as iterators and generators). However, the interpreting philosophy described above is maintained. Moreover, some built-in data structures and procedures are developed in C, in order to increase performance, and Python just offers an interface to them.

7.2 The Python Environment

The source code shown below gives a first example of a Python script.

```
v = input('please input a value')
print('the input value is', v)
```

The script reads a value from the standard input, by means of the `input` function, and assigns such a value to the variable `v`. The assigned value is printed on the screen by means of the function `print`.

The Python 3 interpreter can be called by typing `python3` in a console environment, after that the Python 3 interpreter has been installed in the system. If no input file is given to the `python3` command, an interactive python console is open and the user is allowed to input the instructions manually line by line. Each line, or block of lines, is immediately executed after its insertion. The following listing shows as the previous example runs in the interactive console. Lines that are the source code of the program are prefixed by >>>, and lines that correspond to the interaction between the environment and the user are listed without a particular prefix.

```
>>>v = input('please input a value')
5
>>>print('the input value is', v)
the input value is 5
```

Of course, the use of an interactive console may result helpful for *on-the-fly* calculations or simple analyses that use a pre-programmed procedure. However, developers will find the possibility to write scripts, save them on files and run those files a second time, which is more practical and reusable. Thus, source files can be passed as input to the Python interpreter by specifying them as arguments to the `python3` command. For example, if the source code is stored in a file named `myscript.py`, it can be executed by typing `python3 myscript.py` in the same directory where the file resides. This emphasizes the fact that what the Python interpreter does on reading an input script is exactly the same as manually typing the lines of the script in the interactive console, in the order in which they are written into the file.

Built-in types and variables

The built-in data types that are available in Python 3 are listed in the following table.

Name	Description
`int`	Integer number
`float`	Floating point number
`bool`	Boolean value
`complex`	Complex number with real and imaginary parts
`str`	String
`byte`	Representation of a binary encoding of a string
`list`	A mutable ordered list of values
`tuple`	An immutable tuple of values
`set`	A mutable set of values
`frozenset`	A immutable set of values
`dict`	A mutable key-value map

Primitive types are `int`, `float`, `bool`, `complex`, `str` and `byte`, while the other built-in types are structures of data. Some languages, such as Java, define a clear difference between primitive types and structured objects, and only the latter can be equipped with functions that do not just concern with the storage of the values. In this sense, every Python data is an object which provides both storage and additional functions. For example, `int` is a basic object which also provided the function `bit_length` for calculating the number of bits necessary to represent the specific integer value in binary.

Built-in data types are also divided into **immutable** and **mutable** types. Primitive types, `tuple` and `frozenset` are immutable, while types `list`, `set`, and `dict` are mutable. In Python, there is no general notion of array or matrix, which is usually implemented via lists and lists of lists. Lists are mutable, thus they are inefficient with respect to a *classical* array data structure, but `tuple` and `frozenset` overcome efficiency limitations at the cost of immutability.

There are other data types that are not built-in but they are available via modules of the standard Python library. `Decimal` that is provided within the `decimal` module, and gives a more accurate representation of decimal numbers that overcomes some limitations of the floating point numbers. `fraction`, within the module `Fraction`, allows to represent fractions and provides some operators for them. The module `collection` provides several data structures for collecting objects with specific aims, such as `namedtuple` for representing tuples in which the elements are indexed by a key that can be not only an integer (elements in simple tuples can only be accessed by integer keys that are their position within the tuple), `OrderedDict` that are maps in which keys are stored in an ordered way (the simple key-value structure is implemented via hash tables thus it does not ensure an order), `defaultdict` that are maps in which a default value is specified (simple maps raise an error in case of a missing key) and `Counter` which are maps where the values are the corresponding number of times the keys are inserted.

A **variable** is an object assuming *values* that can change according to situations influencing the object. A name, or label, is associated to a variable along the source code of a program. This notion can be further extended, by considering a second level of variability, with the concept of *reference variable*, where a variable, often called *identifier*, assumes as value a variable having its own value. In other words, two associations are combined: *identifier* $\mapsto$ *variable* $\mapsto$ *value*, where not only *value*, but also *variable* can vary. The term variable is usually reserved for a one-level variable, and *non-reference variable* is used for remarking a simple mechanism of variability. Another way of expressing the difference between non-reference and reference variables is by means of the expression *direct variability* for non-reference variables, and the expression *indirect variability* for reference variables. The advantages of indirect variability are very important and provide computational power that can be appreciated in the development of complex programming.

A physical model of non-reference and reference variables is given by an *addressable memory*, that is, a set of *locations*, each of them identified by a number, called *address*. Locations can contain values, possibly of different types and aggregated in some way. In particular, a location L can contain the address i of a location L'. In this case, we say that i is a **pointer** to L'. This mechanism makes it possible to realise a two-level variability: *identifier* $\mapsto$ *variable* $\mapsto$ *value*. In fact, L plays the role of the identifier, the location L', to which L points by means of i, plays the role of the variable, and the content $\bar{L}'$ of L' the role of the value of L' and of the *referenced value* of L (actually the value of L is L'). In the association chain: $L \mapsto L' \mapsto \bar{L}'$ the pair $L \mapsto L'$ is also called an *environment*, while $L' \mapsto \bar{L}'$ is called a *state*, moreover, L' is the *reference* of L, while $\bar{L}'$, the *direct value* of L', is the *indirect (referenced) value* of L.

In general terms, the *global state* of any computation is given by the associations that, at many possible levels, are established between variables and values. The execution of a computation is essentially a transformation of this global state. This crucial role of variables in computations is the basis of the *Object Oriented Programming* (OOP) paradigm, being objects a generalization and abstraction of variables.

Python treats variables similarly to Java. Thus, a variable of a primitive data type is a non-reference variable, while a variable of a non-primitive type is variable by reference. In this sense, if v and w are variables of a primitive type, then an equation such as $v = w$ is *asymmetric* because it has to be read from the right to the left, as expressing that the value of w is associated to variable v, we say this equation an *assignment* (in some languages an asymmetric symbol like := is used in this case). However, an equation $v = w$ over non-primitive types means that the two reference variables point to the same object. In this case, when the value of v is modified also the value of w is modified too (and *vice versa*).

Differently from Java, a Python variable does not need to be declared and its type is inferred when a value is assigned to it. The actual type of an object can be obtained by the `type` function.

```
x = 5
print( type(5) )
```

```
<class 'int'>
```

The following example shows how to create variables that represent primitive data types. Comment lines start with the character #, and can also be inserted at the end of an instruction line.

```
# this is an integer value
i = -516
# this is a float value
f = 3.14
# this is a boolean value
b1 = True
b2 = False # this is another boolean value
# this is a sequence of byte
y = b'\x00\x01'
z = b'Python' # this is another sequence of byte
# this is a string
s1 = 'my string 1'
s2 = "my string 2" # this is another string
```

The following two examples show the difference between a reference and a non-reference variable.

Variables representing primitive data types directly correspond to the data itself and are non-reference variables. Thus, an assignment `i = j` determines the value of the variable `i` as the value of variable `j`. In this way, a change to the value of `j` does not alter the value of `i`.

```
i = 512
j = i
j = 516
print(i)
print(j)
```

```
512
516
```

In the case of non-primitive data types, such as a list, variables are reference variables, and an assignment of a reference variable such as `l2 = l1`, does not assign a list value to `l2`, but instead, `l2` is a new reference to the value of `l1`. In this way, any modification made to one of the two variables affects the value to which both refer.

```
l1 = list([1,2,3])
l2 = l1
l2.remove(1)
print(l1)
print(l2)
```

```
[2, 3]
[2, 3]
```

7.3 Operators

Operators provide results associated, by some specific rule, with some elements taken as arguments. For example, arithmetic operators correspond to the usual operations on numbers (in the standard infix notation with the operator between the arguments): addition +, subtraction -, multiplication `*`, division / and modulus `%`, exponentiation `**`, and floor division `//` (which returns only the integer part of a division).

Bitwise operators express operations on the bits that compose the numbers. Binary AND &, binary OR | and binary XOR + return a number where each bit is set according to the results of the binary operator. The NOT operator inverts all the bits of a number (`x = y`). The bit shift operators push zeros from one side of the series of bits of a number and shift them on the other side, « and ».

Arithmetic and bitwise operators can be combined with assignment statements, such as `x += 1` or `x &= y`. Python does not provide increment or decrement operators such as ++ and -.

Logical operators are provided via the keywords `and`, `or` and `not`. They can operate on heterogeneous operands by automatically converting the value of the operands. Zero values and empty strings are considered `False` and any other value or string is `True`

Classical comparison operators are provided for working with numbers that return a boolean value (`True` or `False`), such as equality ==, inequality `!=`, greater than >, less than <, greater than or equal to >= and less than or equal to <=. Comparison operators can also be applied to strings. In this case, the comparison takes into account the lexicographic ordering of the two string operand for calculating the result of the comparison.

String operators are the concatenation + (`'a' + 'b' = 'ab'`) and the repetition * (`'ab' * 2 = 'abab'`). The `len` built-it function returns the length of a string (`len('abc')`). In addition, strings are supported by the slice operator for extracting subsequences. It is provided in the form `[start:end:step]`, where the start is the (inclusive) position from which to `start` the slicing, `end` is the (non-inclusive) position at which the slicing is stopped, and `step` defines the number of position shifts to extract the next character.

Indentation

Standards and guidelines for what concerns good programming practices have been released by the Python Software Foundation in a document called PEP (Python Enhanced Proposal). One of the main differences between Python and any other programming language regards code indentation. In fact, contrary to the most used syntax for dividing sections of source code that are based on brackets or *do-done* statements, the Python language uses indentation levels to group sections of the source code. For example, the body of a function is defined by an increased indentation level after the declaration of the function. Both spaces and tabulation characters can be used for the indentation. The PEP 8 documents suggest the use of 4 spaces for each indentation level. Moreover, it discourages the use of tabulation characters and the mixing of spaces and tabulations.

7.4 Statements

Python provides most of the classical control flow statements of programming languages, such as conditional and looping statements. For what concerns conditional statements, Python provides the *if-elif-else* structure. The *if-elif-else* statement if defined as follow, where the *elif* and *else* parts are optional:

```
if <boolean condition>:
    <code block>
elif <boolean condition>:
    <code block>
else:
    <code block>
```

The `if`, `elif` and `else` keywords must be at the same indentation level, as well as the internal lines of the code blocks. Of course, internal code blocks may in turn be structured and contain nested statements.

```
x = 5
if x > 0:
    print('the number is positive')
    if x > 100:
        print('and it is greater than 100')
elif x < 0:
    print('the number is negative')
else:
    print('the number is exactly zero')
```

Python does not provide the *switch-case* statement, but it can be implemented via a series of `elif` conditions.

```
x = input('choose a single character')
if x == 'a':
    print('the character is a')
elif x == 'b':
    print('the character is b')
elif x == 'c':
    print('the character is c')
elif x == 'd':
    print('the character is d')
else:
    print('the character is something else')
```

For what concerns the looping statements, Python provides the *while* and the *for* statements. The *while* statement is defined as follows:

```
while <positive boolean condition>:
    <code block>
else:
    <code block>
```

The `else` statement is optional and it is executed as soon as the while condition becomes `False`. The following example shows the iteration obtained b increasing the variable `i` from zero until it reaches the value 3.

```
i = 0
while i < 3:
    print(i)
    i = i + 1
else:
    print('finish')
```

```
0
1
2
finish
```

Python also provides the `break` command for breaking looping statements. As in any other programming language, the break affects the nearest looping statement.

```
i = 0
while i < 3:
    if i == 2:
        break
    print(i)
    i = i + 1
else:
    print('finish')
```

```
0
1
```

Lastly, the *for* statements is defined as follows, where the `else` part is optional:

```
for <object> in <iterable objects>:
    <code block>
else:
    <code block>
```

An iterable object is any kind of object for which its content can be iterated in a certain way. Built-in collections are iterable objects but also strings and sequences are iterable. The following example shows how to iterate and print the characters of the string `Python` from left to right.

```
for c in 'Python':
    print(c)
```

Strictly related to the `for` statement is the concept of generator. A generator is determined by a set of values obtained as a result of a `for` statement. A typical example of generators is provided by the built-in `range` operator which is used to generate numbers in a given range of values, from an initial value to a final value, according to an increment/decrement step. In the most simple form, it generates values from zero to a given number (excluded).

```
for i in range(3):
    print(i)
```

```
0
1
2
```

The following generator starts from the value 1 and generates the successive values by increasing the current value by 2. It stops when the generated value is greater than the maximum wanted value of 10.

```
# generates number from 1 to 10 by increasing of a step of 2
for i in range(1,10,2):
    print(i)
```

```
1
3
5
7
9
```

The concept of iterator, which is an extension of that one of the generator, will be treated in more details1 in Sect. 7.11.

7.5 Functions

A function identifies a sequence of statements. Any function has a name. When the name of the function is given to the Python interpreter, that is, when the function is *called*, then all the statements of the function are executed. Some variables occurring in the statements of the function can be identified as *parameters* of the function (they are listed between parentheses after the name of the function). In this case, when the function is called with some values of its parameters as arguments, then the statements of the function are executed with the parameters having the values corresponding to the arguments of the called function.

Python functions are defined via the `def` keyword. It is required to specify the name of the function and the parameters that it takes as input between parentheses. The definition of the body of the function follows the indentation rules described in Sect. 7.3, thus the instructions that are at the initial level of the body must have the same level of indentation.

```
def print_hello():
    print('Hello!')

print_hello()
```

```
Hello!
```

Parameters are defined by only specifying their names, and not their types.

```
def print_value(x):
    print('This is the value:', x)

v = 'my value'
print_value(v)
```

```
This is the value: my value
```

The scope of parameters and of any variable declared within the body of the function is locally limited to the function itself. However, the function can access variables occurring in an upper level of the source code of which the function is part.

The definition of the function does not include any information regarding the return value. If a return statement is reached by an instruction of the function, then the specified value is returned and the function is terminated.

```
def increase(x):
    return x + 1

v = 0
v = increase(v)
print(v)
```

```
1
```

None, one or multiple values can be returned and they are assigned to variables by following the order in which they are specified in the return statement.

```
def divide(x, y):
    return x // y , x % 2

quotient, remainder = divide(15, 2)
print(quotient, 'remainder', remainder)
```

```
7 remainder 1
```

Python allows the declaration of a default value for input parameters, by specifying it in the function declaration. If a value for the parameter is not specified during the call to the function, the default value is used. Otherwise, the default value is overwritten by the specified one.

```
def pow(x, exponent = 1):
    return x ** exponent

print( pow(2) )
print( pow(2,2) )
```

```
1
4
```

As a consequence, parameters for which a default value is declared become optional. Thus, in some circumstances, it is necessary to specify which value is assigned to which parameters during the call of the function.

```
def do_something(x = 1, y = 2, z = 3):
    return (x + y) * z   # by default (1 + 2) * 3 = 9

print( do_something(4) )  # ?
print( do_something(y = 4) )
```

```
18        # the first optional value is overwritten
15        # x = 1, z = 3 by default values
```

It is a good programming practice to document functions by explaining what the input parameters are, what the function computes, or generally does, and what the function returns. Comment lines, which are generally read by users who are developing the specific function, are useful for understanding the functioning of a piece of code but they do not provide insights to end users. It is allowed to insert special comments, called *docstrings*, to provide end-user documentation. Docstrings are multi-line comments that start and end with three consecutive double quotes. When applied to functions they must be inserted immediately after the function definition and with the same indentation level as the function body. Once specified, the docstring is registered by the python environment as a documentation of the function that can be retrieved by the `help` built-in method.

```
def divide(x, y):
    """
    This function computes the quotient and remainder of a
    division.
    """
    return x // y , x % 2

help(divide)
```

```
Help on function divide in module __main__:

divide(x, y)
    This function computes the quotient and remainder of a
    division.
```

A best practice is to specify, within the docstring of a function, a summary of the function, the type and the meaning of the function parameters and of the returned values.

```
def divide(x, y):
    """
    This function computes the quotient and the remainder of the
    division x / y.

```

```
    ----------
    Parameters
        x (numeric) : the dividend
        y (numeric) : the divisor

    ----------
    Returns
       int : the quotient
       int : the remainder
    """
    return x // y , x % 2
```

7.6 Collections

The Python built-in data collections are `list`, `set` and `dict`. They can contain several primitive data of heterogeneous types, and they can even contain collections themselves. The following source code shows how to define collections for a predefined set of objects. Lists are defined by means of square brackets, while sets are defined with curly brackets, by listing the objects that are contained in the collection and separating them with commas. For avoiding confusion with the genomic dictionaries, we call "maps" the objects of type `dict`, usually called dictionaries. Maps are also defined with curly brackets, but their definition must contain key-value associations that are listed in the form `key : value`.

```
mylist = [1, 's', 'a string']
myset = {1, 's', 'a string'}
mydict = {0: 1, 1: 's', 'x': 'a string'}

print(mylist)
print(myset)
print(mydict)
```

```
[1, 's', 'a string']
{1, 'a string', 's'}
{0: 1, 'x': 'a string', 1: 1}
```

Since, sets and maps do not ensure the ordering of the elements that they store, the print of such data structures does not respect the order in which their elements have been declared.

Collections can also be declared as empty containers, and items can be added successively. Empty list is constructed by `list()` or `[]`.

The following code shows the main functions for adding items to a list, where `append(2)`, `append(3)`, `append(4)` are *methods* because are applied, writing them on the right after a dot, to the variable `mylist` by changing its value. The notion of method extends to any object as it will be explained in the section devoted to classes and objects.

```
# initialize an empty list
mylist = list()

# append items to the tail of the list
mylist.append(2)
mylist.append(3)
mylist.append(4)
print(mylist)
```

```
[2, 3, 4]
```

```
# insert an item at a specific position,
# the first position is indexed by 0
mylist.insert('1', 0)
print(mylist)
```

```
[1, 2, 3, 4]
```

```
# append all the items of a list to the tail of my list
mylist.append( [5,6,7] )
print(mylist)
```

```
[1, 2, 3, 4, 5, 6, 7]
```

Two main methods provide the removal of items from a list and one for entirely cleaning the collection.

```
# initialize a list
mylist = [1, 2, 2, 3]

# remove the first occurrence of a given item
mylist.remove(2)
print(mylist)
```

```
[1 ,2, 3]
```

```
# remove the item at a specific position, and return it
e = mylist.pop(1)
print(mylist)
print(e)
```

```
[1, 3]
2
```

```
# if no index is given, that it removes the last item
mylist.pop()
print(mylist)
print(e)
```

```
[1]
```

```
# clear the list
list.clear()
```

```
3
```

Also, operators are defined for lists. Similarly to strings, the concatenation operator +, and the repetition operator * can be applied. Their usage is intuitive as it is shown in the following example.

```
# initialize a list by concatenating two lists
mylist = [1, 2, 3] + [4, 5]
print(mylist)

# create a list that repeats 3 times a initial list
mylist = [1, 2] * 3
print(mylist)
```

```
[1, 2, 3, 4, 5]
[1, 2, 1, 2, 1, 2]
```

The *slicing* operator is defined in order to retrieve specific portions of lists. It is defined in the form `[start : end : step]` such that the elements going from the starting position (included) to the ending position (excluded) are retrieved by applying a specific step. If the step argument is not specified then simply the slice of the list from the given positions is retrieved. Moreover, default values for the start and the end of the slicing are the head and the tail of the list, respectively.

```
l = [0, 1, 2, 3, 4, 5, 6, 7, 8, 9]

# slicing from a star to and end position
print( l[1:5] )

# slicing to and end position
print( l[:5])

# slicing from a start position
print( l[5:] )

# slicing the entire list by a step of 2
print( l[::2] )
```

```
[1, 2, 3, 4, 5]
[0, 1, 2, 3, 4]
[5, 6, 7, 8, 9]
[0, 2, 4, 6, 8]
```

The square bracket operator is also used to retrieve the item at a given position of the list by considering the first position as having index 0. Moreover, negative values can be specified for the argument of the operator. When negative values are specified for the start and end positions, the list is treated as a circular list and the corresponding positions are calculated by going backwards from the default position. A negative value of the step corresponds to going backwards from the end to the start position by skipping the specified number of positions at each step.

```
mylist = [0, 1, 2, 3, 4, 5, 6, 7, 8, 9]

print( mylist[-1:] )

print( mylist[:-1] )

print( mylist[::-1] )
```

```
[9]
[0, 1, 2, 3, 4, 5, 6, 7, 8]
[9, 8, 7, 6, 5, 4, 3, 2, 1, 0]
```

Lastly, the slicing operator can be combined with the `del` operator to remove from the list the elements selected with the slice.

```
mylist = [0, 1, 2, 3, 4, 5, 6, 7, 8, 9]

# remove the first element
del mylist[0]
print(mylist)

# remove a slice of elements
del mylist[1:2]
print(mylist)

# remove a slice of elements with a step of 2
del mylist[1::2]
print(mylist)
```

```
[1, 2, 3, 4, 5, 6, 7, 8, 9]
[1, 3, 4, 5, 6, 7, 8, 9]
[1, 4, 6, 8]
```

Similarly to lists, sets provides a method for adding elements, that is `add`, and methods for removing items or clearing the collections, namely `remove` and `clear`. However, sets do not prove any functionality that is based on an indexing concept, such as the position of an element.

```
# initialize an empty set
myset = set()

# add an element to a set
myset.add(1)

# adding a duplicated element to a set
# does not affects the content of the set
myset.add(1)

# remove an element form a set
myset.remove(1)
```

Special operators that are defined to be applied to sets are set-theoretic operators, such as union

|, intersection &, difference - and symmetric difference (elements that are in both sets both not in their intersection).

```
# initialize two sets
s1 = {1, 2}
s2 = {2, 3}

# union
print( s1 | s2 )

# intersection
print( s1 & s2 )

# difference
print( s1 - s2 )

# symmetric difference
print( s1 ^ s2)
```

```
{1, 2, 3}
{2}
{1}
{1, 3}
```

Moreover, two operators are defined for checking the inclusion of a set within another set, that are >= and <=.

```
# initialize two sets
s1 = {1, 2}
s2 = {1, 2, 3}

# check is s1 is subset of s2
print( s1 <= s2 )

# check is s1 is superset of s2
print( s1 >= s2 )
```

```
True
False
```

Python provides an easy way to convert lists to sets and vice versa such that the specific methods of both collections can be applied in several situations. For example, sets can be used to remove duplicated elements from a list, by simply converting a list to a set.

```
mylist = [1, 2, 2, 3]

# covert to a set fro removing duplicates
myset = set(mylist)

# convert back to a list
print( list(myset) )
```

```
[1, 2, 3]
```

Lists and sets are iterable objects, thus content can be iterated in a for statement.

```
mylist = [1, 2, 2, 3]

for v in mylist:
    print( str(v) + ', ', end='')
# end='' prevents the new line at the end of the print

print() # make a new line

for v in set(mylist):
    print( str(v) + ', ', end='')
```

```
1, 2, 2, 3,
3, 1, 2,
```

As introduced before, maps are collections of key-value pairs. Keys are unique and each key has only one assigned value. There is no limitation on the type of keys and values, and they can be of different types within the same map. The square bracket operator allows one to assign a value to a key and retrieve the value of a given key.

```
# initialize an empty map
mydict = dict()

# assign 'one' to the integer key 1
mydict[1] = 'one'

# assign 2 to the string key 'two'
mydict['two'] = 2

# print the entire map
print( mydict )

# print the value associated to the key 1
print( mydict[1] )
```

```
{1:'one', 'two':2}
'one'
```

Maps are iterable objects, thus their content can be iterated within a `for` loop. The iteration of a map can be performed in three different ways. The entire set of keys can be iterated, as well as the entire list of values, or the maps can be iterated as a set of key-value items. None of these three methodologies provided an ordered iteration. The following example shows how to use the method `keys` to retrieve the complete set of keys of a map and to print the entire content of the map with a `for` loop.

```
mydict = {1:'one', 'two':2, 3:'three'}

# iterating by the set of keys
for k in mydict.keys():
    print( k, mydict[k])
```

```
1    'one'
'two'    2
3    'three'
```

Alternatively, the method `items` can be used to retrieve the complete set of key-value pairs contained in a map. These kinds of items are represented as tuples, thus the iterating object is a tuple with two values, the first is the key and the second is the assigned value.

```
mydict = {1:'one', 'two':2, 3:'three'}

# iterating by the set of key-value pairs
for t in mydict.keys():
    print(t, t[0], t[1])
```

```
(1,'one')    1    'one'
('two',2)    'two'    2
(3,'three')    3    'three'
```

Python provides a special behaviour for decomposing tuples into distinct values. The iteration of a map is a typical example. The item tuple is decomposed into two values `k` and `v`, which name is defined by the developer.

```
mydict = {1:'one', 'two':2, 3:'three'}

# iterating by the set of key-value pairs
for k,v in mydict.items():
    print( k, v )
```

```
1    'one'
'two'    2
3    'three'
```

For what concerns the retrieving operation, if the key is not present in the map then an error arises. Thus, the existence of the key must be verified before every access to the map that is not an assignment. The existence can be verified with the `in` operator which returns true in case the item is an existing key of the map.

```
# initialize an empty map
mydict = {1:'one', 'two':2}

if 3 in mydict:
    print( mydict[3] )
else:
    print('key not found')
```

```
key not found
```

The `in` operator can also be combined with the logical `not` operator.

```
# initialize an empty map
mydict = {1:'one', 'two':2}

if 3 not in mydict:
```

```
    mydict[3] = 'three'
else:
    print('a value already exists for the key 3')

print(mydict)
```

```
{1:'one', 'two':2, 3:'three'}
```

The method `get` provided by the `dict` class is an alternative to the use of the `in` operator. The method has the same behaviour as the square bracket operator for retrieving the value assigned to a key, however, a default value can be specified in case of a missing key such that the default value is returned and no error arises.

```
# initialize an empty map
mydict = {1:'one', 'two':2}

print( mydict.get(3, 'not found') )
```

```
not found
```

The `in` operator can also be used for lists and sets to check the existence of an item within the collection.

```
mylist = [1, 2, 3]
print( 4 in mylist )
```

```
False
```

Comprehension is an instrument that allows the creation of collections in a concise way by starting from some generated values. Comprehension also allows us filtering and transforming the given element before it is inserted into the collection.

For what concerns lists, the comprehension is in the form

```
[ <transformation> <generative loops> <filters> ]
```

Initial values are generated in the middle block of the form, which consists of one or more generative `for` loops. Filters are applied to generated values, and transformations are finally applied in order to obtain the resultant values. The transformed values compose the created list.

For example, a list which contains the numbers from 1 to 10 is created in the following way:

```
[ i for i in range(1,11) ]
```

```
[1, 2, 3, 4, 5, 6, 7, 8, 9, 10]
```

In this case, no filters and transformations are applied. The `for` statement generated the values iteratively, and the iteration variable `i` is used to insert values within the collection.

Comprehension with filtering can be used to generate numbers multiples of 2 form 2 to 10:

```
[i for i in range(2,10) if (i % 2) == 0 ]
```

```
[2, 4, 6, 8, 10]
```

In this case, the filter `if (i % 2) == 0` is applied to the iteration variable, and the values of the iteration that pass the filter are inserted into the resultant list.

The next example shows how to apply a transformation in order to obtain the powers of 2 from 2^0 up to 2^6 (we recall that `**` is the power operator).

```
[ 2**i for i in range(0,7) ]
```

```
[1, 2, 4, 8, 16, 32, 64]
```

In the same way, comprehension can be used to generate sets and maps. Comprehension for sets is equal in almost everything to the comprehension for lists except that it is encapsulated within curly brackets instead of square brackets. The following examples show how to create a set that contains the squares of numbers from 1^2 to 10^2 put together with the number themselves in a pair:

```
{ (i, i*i) for i in range(1,11) }
{ (i, i*i) for i in range(1,11)}
```

```
{(1, 1), (2, 4), (3, 9), (4, 16), (5, 25), (6, 36), (7, 49),
 (8, 64), (9, 81), (10, 100)}
```

In this case, the generated value is not just a number but it is a more complex item, such as a tuple.

Comprehension can also be applied to generating maps. The previous example is converted into a map where keys are the numbers from 1 to 10 and values are their square:

```
{ i : i*i for i in range(1,11) }
```

```
{1: 1, 2: 4, 3: 9, 4: 16, 5: 25, 6: 36, 7: 49, 8: 64, 9: 81,
10: 100}
```

Lastly, the function `len` is applied to any type of collection in order to retrieve the number of elements that are stored in the data structure. In the case of maps, the built-in function returns the number of key-value items that are currently inside the map.

```
mylist = [i for i in range(1,11)]
print( len(mylist) )
```

```
10
```

7.7 Sorting

The built-in operator `sorted` sorts input lists of sets and returns an ordered list.

```
myset = {2,1,3}
print( sorted(myset) )
```

```
[1,2,3]
```

It is required that the elements of the collection must be comparable to each other. Thus, the sorting function is not able to compare a numeric with a non-numeric object, for example, an integer with a string. An error of type `TypeError` is thrown if incomparable types are sorted.

```
myset = {2,1,'a string'}
print(myset)
```

```
------------------------------------------------------------------------
ValueError                              Traceback (most recent call last)
<ipython-input-27-45c0c9bb2cec> in <module>()
----> 1 print(sorted(myset))

TypeError: unorderable types: str() < int()
```

Boolean values are exceptions because they can be interpreted also as numbers or strings. In fact, the integer values of `True` and `False` are 1 and 0, respectively, and their string representation simply reports their keywords.

```
print(int(True), int(False))
print(str(True), str(False))
```

```
1 0
True False
```

Notably, the `None` constant can not be converted to a numeric value but only to literal and boolean values.

The `sort` function receives two optional arguments, `reverse` and `key`. The `reverse` argument specifies that the sorting must be performed in a reverse ordering, namely from the highest to the lowest value.

```
sorted( [4,2,1,3] )
sorted( [4,2,1,3], reverse = True )
```

```
[1, 2, 3, 4]
[4, 3, 2, 1]
```

The `key` argument specifies a function (function will be treated in Sect. 7.5) that is applied to the elements of the collection and is used to give a representation of the elements that will be used for the sorting. For example, the comparison between two tuples iteratively takes into account the values of the tuples from the left-most to the right-most. The first comparison that is not an equality result gives the ordering of the two tuples. In the following example, the two tuples are `(1,3)` and `(2,0)` and the resultant order is computed by only taking into account the values 1 and 2.

```
sorted( [(1,3), (2,0)] )
```

```
[(1,3), (2,0)]
```

The sorting of the two tuples by their second values can be obtained by using the `key` argument in the following way:

```
def get_second(t):
    return t[1]

sorted( [(1,3), (2,0)], key=get_second )
```

```
[(2,0), (1,3)]
```

In this case, the function `get_second` is applied to each element of the list in order to extract an alternative value that is representative of the element for its comparison with the others.

The default toolkit of Python provides the function `itemgetter` available by the `operator` module. that is an alternative to the writing of a new function for the developer. The function takes as a parameter the index of the value within the tuple that must be used for the comparison.

```
from operator import itemgetter

sorted( [(1,3), (2,0)], key=itemgetter(1))
```

```
[(2,0), (1,3)]
```

7.8 Classes and Objects

This section gives a brief introduction to object-oriented programming (OOP) in Python by focusing on how to define classes. We recall that in the OOP paradigm, an object is an instance of a given class. A class definition specifies how to construct and modify the **objects** of the class. The concept of class has a hierarchical organization of classes and subclasses.

An object of a class is constructed by a *constructor* identified by a reserved name given by `__init__`. Any constructor must be declared with an initial parameter that is called `self`. This parameter is a reference to the object and it is used to modify the object by obtaining another instance of the object. Therefore, different objects of a class correspond to instances of the class, while different instances of an object correspond to different states of the object, conceived as a dynamical entity changing during computations. In the following example, the definition of the class `Coordinate` describes objects that represent Cartesian coordinates.

```
class Coordinate:
    """
    A class to represent Cartesian coordinates
```

```
    """

    x = 0
    # The x coordinate. By default it is 0

    y = 0
    # The y coordinate. By default it is 0

    def __init__(self, x_coord, y_coord):
        self.x = x_coord
        self.y = y_coord

c1 = Coordinate(2,3)
c2 = Coordinate(2,4)
```

The example shows the definition of the class and the creation of two instances, `c1` and `c2`, that represent two distinct points in the Cartesian space.

Parameters inside parentheses of `__init__` are called *attributes* of the objects. The attributes are by default publicly accessible. They can be retrieved by the dot `.` operator applied to an instance of the class. In the following example, the coordinates of one point are read.

```
c1 = Coordinate(2,3)
print( c1.x, c2.y )
```

Since the access is public, the internal attributes of the class can also be written by other entities.

```
c1 = Coordinate(2,3)
c1.x = 1
print( c1.x, c2.y )
```

```
1 3
```

However, the access to the internal state of an object by means of its attributes violates good programming practices in OOP, a principle called *encapsulation*.

7.9 Methods

In order to implement the encapsulation, access to the class attributes must be forbidden and specific functions for reading and writing the values of attributes must be equipped within the class definition. These functions are called **methods** of the class and are applied to the objects of the class by means of the standard dot-suffix notation. In Python syntax, any attribute whose name starts with a double underscore is a private variable, thus it cannot be externally accessed.

```
class Coordinate:
    """
    A class to represent Cartesian coordinates
    """
```

```python

    __x = 0
    # The x coordinate. By default it is 0

    __y = 0
    # The y coordinate. By default it is 0

    def __init__(self, x_coord, y_coord):
        self.set_x(x_coord)
        self.set_y(y_coord)

    def get_x(self):
        return self.__x

    def get_y(self):
        return self.__y

    def set_x(self, x):
        if x >= 0:
            self.__x = x

    def set_y(self, y):
        if y >= 0:
            self.__y = y

c = Coordinate(2,3)

# print( c.__x )   # __x is private, this will throw an error

print( c.get_x() )

c.set_x(1)
```

As standard guidelines for OOP, get and set methods are implemented for accessing the internal state in a controlled way. In fact, the encapsulation allows the class to control how the internal state is modified, and this helps the object to ensure a consistent form of its state. Thus, only internal mechanisms that are relevant for using the object are made available to external entities. In the example, the class is implemented such that it works only with positive coordinates, and if negative values are specified then they are ignored.

In modern languages, objects are equipped with special functions for obtaining a printable description of them. Python does not make an exception to this practice. The method is called `__repr__`, and it is used by the Python environment every time it tries to cast an instance of the class to a string.

```python
class Coordinate:
    __x = 0
    __y = 0
    #class_body
    def __repr__(self):
        return '('+str(self.__x)+','+str(self.__y)+')'

```

```
c = Coordinate(2,3)
print(c)
```

```
(2, 3)
```

More in general, Python reserves some special names to be assigned to functions for implementing specific behaviours. These names are recognizable since they start and end with a double underscore.

Other interesting special methods are those reserved for defining how Python operators work on custom objects. This concept is usually called operator overloading. Each operator has a corresponding function name. For example, the + operator is defined via the `__add__` function. Special methods that implement operators must ensure that the operand belongs to the specific class. This check is performed by the statement `assert` in conjunction with the built-in function `isinstance`. The built-in function returns a positive value if the object is an instance of a given class. The assertion will throw an error in case of a negative result. The addition operator works on two operands. The left operand is identified by the `self` instance, and the right operand is identified by the additional function parameter.

```
class Coordinate:
    __x = 0
    __y = 0
    #class_body
    def __add__(self,other):
        assert isinstance(other, Coordinate)
        return new Coordinate(self__x + other.__x,
                              self__y + other.__y )

    def __sub__(self,other):
        assert isinstance(other, Coordinate)
        return new Coordinate(self__x - other.__x,
                              self__y - other.__y )

c1 = Coordinate(2,4)
c2 = Coordinate(2,3)

print( c1 + c2 )
print( c1 — c2 )
```

```
(4,7)
(0,1)
```

A set of interesting operator overloading functions are those regarding comparison operators. In the following example, a function to compare coordinates is first developed, and then it is used for implementing the comparison operator. The function, that in the following example is called `compare`, and follows some good practices in programming. It takes as input the current instance and an object with which the instance must be compared. If the two entities are considered equal, then the result of the comparison is zero. Otherwise, the result is a negative integer value if the instance is considered less than the compared object, otherwise, it is a positive integer value.

This practice implies the definition of an ordering between objects of the given class. Moreover, it allows the specification of the amount of difference between the objects. For example, the Hamming distance can be used for comparing strings. The distance counts in how many positions the two strings have a different character. Thus, the ordering function adds a sign to the distance for deciding the order between two objects.

In the following example, a comparison between coordinates is defined such that they are firstly compared by the *x* coordinate, and in the case of equality they are compared by the *y* coordinate.

```python
import math

class Coordinate:
    ...
    def compare(self, other):
        assert isinstance(other, Coordinate)
        if self.__x == other.__x:
            return self.__y - other.__y
        return self.__x - other.__x

    # Overloading of comparison operators #
    def __lt__(self, other):
        return self.compare(other) < 0
    def __gt__(self, other):
        return self.compare(other) > 0
    def __eq__(self, other):
        return self.compare(other) == 0
    def __le__(self, other):
        return self.compare(other) <= 0
    def __ge__(self, other):
        return self.compare(other) >= 0
    def __ne__(self, other):
        return self.compare(other) != 0

c1 = Coordinate(2,3)
c2 = Coordinate(2,4)

print( c1 < c2 )
print( c1 >= c2 )
```

```
True
False
```

Python is in many of his aspects a structure-less language. Variables do not need to be declared, for example. Functions are overwritten when multiple declarations with the same function name are given.

```python
def foo()
    print('foo')

foo()

def foo(x = 0):
```

```
    print('foo',0)

foo()
```

```
foo
foo 0
```

Objects are not an exception to this kind of fluidity that is in Python. As said before, objects are in their essence containers of properties. The list of the properties associated with a given object can vary over time. This means that the class declaration is a template that specifies the guidelines for building an initial structure of an object that instantiates such a class. Such a structure can further be modified. The following example shows how it is possible to add an internal variable to an object after its instantiation.

```
class Myclass:
    x = 0
    def __init__(self, x ):
        self.x = x

o = Myclass(1)
o.y = 10
print(o.y)
```

```
10
```

7.10 Some Notes on Efficiency

Python has changed during the time its internal behaviours, especially from version 2 to version 3. The changes do not only regard the syntax of the language but include modifications to the internal behaviour of built-in functions. Python is a very high-level programming language, and this feature gives to it a series of advantages. However, a legitimate developer who takes into account the underlying functioning of the Python environment will take benefits from developing software of good quality and increased performance. The choice of using a specific coding strategy is of main importance.

In this perspective, **generators** provide a typical case study to understand the importance of good coding behaviour. As an example, the developer is asked to write a source code for retrieving all the prime numbers that are smaller than a given number. A trivial solution is to iterate over the number from 1 to the maximum required value and to check for each number if it is a prime number. The following example uses an already developed function, `is_prime`, to perform the check, which implementation is not the focus of this example and thus it is ignored.

```
max_value = 100
prime_numbers = set()

```

```
for i in range(1, max_value):
    if is_prime(i):
        prime_numbers.add(i)
```

The solution is implemented by using the function `range`, which creates a generator for navigating through the numbers that have to be tested. An important fact is that the generator does not create a complete list of numbers from 1 to the maximum value. On the contrary, one number at a time is stored in memory. An alternative would be the use of list comprehension as it is shown in the following source code:

```
max_value = 100
prime_numbers = set()
numbers = [i for in in range(1,max_value)]

for i in numbers:
    if is_prime(i):
        prime_numbers.add(i)
```

The choice of using this strategy has a main pitfall, that is, the creation of a list that is completely stored in memory. If the maximum value is relatively small, such as 100, the issue has a low impact on the performance of the software, but it may be disruptive for high values.

Another example of good programming regards the sorting of a list. In this toy case, the developer has to order a given list of numbers. A non-expert developer will probably sort the list via the built-in function `sorted` which creates a copy of the input list. Such a copy reports elements in an ordered way. Then, the original list must be overwritten with the sorted copy.

```
ilist = [2,4,1,3]
ilist = sorted( ilist )
```

If there is no need for creating a copy of the list in the original ordering, then it should be preferable to not create such a copy. This is of high impact on the performance, especially if the list is relatively large: It takes twice the memory, but it also requires additional time to create the copy to be sorted. Thus, in this situation, it is preferable to use the method `sort` provided by the list data type. The method acts on the original data structure without the need for a creation of a copy.

```
ilist = [2,4,1,3]
ilist.sort()
```

7.11 Iterators

Generators described in the previous section are special kinds of **iterators**, that is, objects identifying a collection, but without storing all the elements of the collection, but only one of them at each time. In this way, a gain of space is realized at the cost of a computational mechanism of updating objects along the computation, by elaborating on each of them, at any step of an iterative process.

Two important steps of an iterator are the initialization and the passing from one item to another. In this perspective, an iterator is an entity which maintains an internal state. The state is initialised in a given way, and special procedures define how to pass from one state to another. If the iterator is in a consistent state, then it can be asked for the value that corresponds to its internal state.

In Python, iterators can be implemented in several ways. We focus on defining iterators as objects. Iterators are special types of objects which implement the `__iter__` and `__next__` methods.

The following example illustrates how to build an iterator that iterates over the integer numbers starting from 1. The internal state is represented by the internal variable `__i`. Each time the method `next` is called on the iterator, the internal state is updated with a forward operation and the element resulting from such an operation is returned.

```
class my_iterator:
    __i = 0

    def __init__(self):
        self.__i = 0

    def __next__(self):
        self.__i += 1
        return self.__i

it = my_iterator()
print( next(it) )
print(next(it))
```

```
1
2
```

One important thing to remember is that the method `next` is called in order to retrieve the elements of the iteration. This means that, after the creation of the iterator, the first element is retrieved only after a call of the `next` method. For this reason, in order to start the iteration from the value 1, the initial state is set to 0.

The given class is still not suitable to provide objects that can be used in `for` statements. To this end, an explicit definition of `__iter__` method is necessary that makes explicit the iterable nature of the defined objects, as it is shown in the following class definition.

```
class my_iterator:
    __i = 0

    def __init__(self):
        self.__i = 0

    def __iter__(self):
        return self
```

```
    def __next__(self):
        self.__i += 1
        return self.__i

it = my_iterator()
for n in it:
    print(n)
    if n > 10:
        break
```

```
1
2
3
4
5
6
7
8
9
10
11
```

The object provides an iteration over integer numbers without any limit. In fact, it can iterate up to infinite. The method `__next__` has to raise a `StopIteration` exception in order to stop the interaction at a specific state. For example, an iterator to enumerate integers from 1 to 10 is the following one.

```
class my_iterator:
    __i = 0
    __limit = 10

    def __init__(self):
        self.__i = 0

    def __iter__(self):
        return self

    def __next__(self):
        self.__i += 1
        if self.__i <= self.__limit:
            return self.__i
        else:
            raise StopIteration

it = my_iterator()
for n in it:
    print(n)
```

```
1
2
3
4
```

```
5
6
7
8
9
10
```

The raise of a `StopIteraction` exception is a special event that, in the case of iterator objects, is not threaded like a common exception raise. In fact, it is not cached by the try-except statement.

```
it = my_iterator()
try:
    for n in it:
        print(n)
except:
    print('an exception has been arised')
```

```
1
2
3
4
5
6
7
8
9
10
```

```
def m_function():
    raise StopIteration

try:
    my_function()
except:
    print('an exception has been arised')
```

```
an exception has been arised
```

7.12 Itertools

Python provides a set of built-in functions to build combinatorial structures. The library is module `itertool`. and its documentation is given at https://docs.python.org/3/library/itertools.html. It can be used, for example, to perform Cartesian products.

```
import itertools

alphabet = 'acgt'
```

```

# cartesian product, equivalent to a nested for–loop
itertools.product(alphabet, repeat = 2)

for i in itertools.product(alphabet, repeat = 2):
    print(i)
```

```
('a', 'a')
('a', 'c')
('a', 'g')
('a', 't')
('c', 'a')
('c', 'c')
('c', 'g')
('c', 't')
('g', 'a')
('g', 'c')
('g', 'g')
('g', 't')
('t', 'a')
('t', 'c')
('t', 'g')
('t', 't')
```

The conversion to usual Python strings can be obtained with the following for statement.

```
for i in itertools.product(alphabet, repeat = 2):
    print( ''.join(i) )
```

```
aa
ac
ag
at
ca
cc
cg
ct
ga
gc
gg
gt
ta
tc
tg
tt
```

And many other combinatorial structures can be generated.

```
for i in itertools.product(alphabet, repeat = 2):
    print(''.join(i), end=' ')
print()

# r–length tuples, no repeated elements
```

```
for i in itertools.permutations(alphabet, 2):
    print(''.join(i), end=' ')
print()

# r-length tuples, in sorted order, no repeated elements
for i in itertools.combinations(alphabet, 2):
    print(''.join(i), end=' ')
print()

# r-length tuples, in sorted order, with repeated elements
for i in itertools.combinations_with_replacement(alphabet,2):
    print(''.join(i), end=' ')
print()
```

```
aa ac ag at ca cc cg ct ga gc gg gt ta tc tg tt
ac ag at ca cg ct ga gc gt ta tc tg
ac ag at cg ct gt
aa ac ag at cc cg ct gg gt tt
```

Chapter 8
Laboratory in Python

8.1 Extraction of Symbols

Given a string *s* we would like to know if it is contained in the genome. We can use the `set` data structure, which is a modifiable collection of elements with no repetitions. We iterate over the positions of *s* and extract the characters at the given position.

```
s = 'agctaggaggatcgccagat'

alphabet = set()
for i in range( len(s) ):
    alphabet.add( s[i] )

print(“your alphabet is: ”, alphabet)
```

The result is:

```
your alphabet is: {'g', 'c', 'a', 't'}
```

Alternatively, since a string is an iterable object, we can iterate over the characters of *s* with no indication of their position.

```
alphabet = set()
for c in s:
    alphabet.add( c )

print(“your alphabet is: ”, alphabet)
```

```
your alphabet is: {'t', 'a', 'c', 'g'}
```

The `set` data structure does not assure any specific order regarding the elements contained in it. Thus multiple runnings of the same source code can generate the same set but print the symbols in a different order. The `sorted` function returns a list of sorted elements.

V. Manca and V. Bonnici, *Infogenomics*, Emergence, Complexity and Computation 48,
https://doi.org/10.1007/978-3-031-44501-9_8

```
print("your sorted alphabet is: ", sorted(alphabet) )
```

```
your sorted alphabet is:  ['a', 'c', 'g', 't']
```

The `sorted` function requires that there must exist an ordering relation between the elements of the input collection. The following instructions generate an error because there is no defined ordering relation between strings (or simple characters) and integers.

```
myset = {'a', 0}
print(sorted(myset))
```

However, the last instruction throws an error:

```
--------------------------------------------------------------------------
TypeError                               Traceback (most recent call last)
<python-input-25-d74947341104> in <module>()
      1 myset = {'a', 0}
----> 2 print(sorted(myset))

TypeError: unorderable types: str() < int()
```

The way lists and sets (but more in general any built-in data structure) are printed is defined by Python. Lists are delimited by square brackets, while sets are delimited by curly brackets, and elements are separated by commas. However, we would like to personalize the print of our alphabet. A first approach could be the creation of an empty string that will be concatenated to the symbols of the alphabet such that they are interleaved by a separator character.

```
alpha_string = ''

# sets are iterable objects, thus they can be used for cycling
for a in sorted(alphabet):
    alpha_string = alpha_string + ',' + a

print('your sorted alphabet is: ' + alpha_string)
```

```
your sorted alphabet is: ,a,c,g,t
```

Since the first comma is unwanted, we can extract the created string without the first character.

```
print('your sorted alphabet is: ' + alpha_string[1:])
```

```
your sorted alphabet is: a,c,g,t
```

The same string can be obtained by using a function defined for the string class, called `join`. Given a string and a collection of elements, the `join`function concatenate the element by putting a copy of the string between them.

```
print( ','.join(alphabet) )
```

```
t,a,c,g
```

We can use the join function on lists generated by comprehension as well.

```
','.join( [ i for i in range(0,11)] )
```

```
---------------------------------------------------------------------------
TypeError                                 Traceback (most recent call last)
<ipython-input-16-475d059feeaf> in <module>()
----> 1 ','.join( [ i for i in range(0,11)] )

TypeError: sequence item 0: expected str instance, int
found
```

However, we need to remember that the concatenation operation is not defined in python between elements that are not strings. Thus, if we want to concatenate elements that are not strings, we need to explicitly cast the elements to be strings.

```
','.join( [ str(i) for i in range(0,11)] )
```

```
'0,1,2,3,4,5,6,7,8,9,10'
```

```
print("your sorted alphabet is: " +
    ','.join( [ str(a) for a in sorted(alphabet)] ) )
```

```
your sorted alphabet is: a,c,g,t
```

8.2 Extraction of Words

The extraction of the alphabet of a string corresponds to the extraction of the 1-mers contained in the string. The procedure can be generalized to extract 2-mers in the following manner:

```
words = set()
for i in range(len(s)):
    words.add( s[i:i+2] )
print( words )
```

```
{'at', 'gg', 'ca', 'ga', 'tc', 'cc', 't', 'ta', 'ag', 'cg', 'ct',
'gc'}
```

A procedure giving the same result can be represented as a sliding window, having a size equal to two, and that starts from the first position of the string, then it extracts the first 2-mer and, subsequently, extracts the other 2-mers by moving forward of one step. An easy way to visualize all the extracted windows is to append them to a list and then print the list:

```
words = list()
for i in range(len(s)):
    words.append( s[i:i+2] )
print(s)
print( words )
```

```
agctaggaggatcgccagat
['ag', 'gc', 'ct', 'ta', 'ag', 'gg', 'ga', 'ag', 'gg', 'ga', 'at',
'tc', 'cg', 'gc', 'cc', 'ca', 'ag', 'ga', 'at', 't']
```

For avoiding the extraction of the last *t*, which is not a 2-mer, a general function `get_kmers` is defined that takes as input the string *s* and the word length *k*, and it returns all and only the k-mers occurring in *s*.

```
def get_kmers(s, k):
    """ This function returns the set of k-mers that occur in a
    given string """

    kmers = set()
    for i in range(len(s) - k +1):
        kmers.add(s[i:i+k])
    return kmers

kmers = get_kmers(s,3)
print(kmers)
```

```
{'gga', 'gag', 'cag', 'tag', 'cgc', 'cta', 'gcc', 'aga', 'tcg',
'agg', 'cca', 'atc', 'gct', 'agc', 'gat'}
```

8.3 Word Multiplicity

In what follows, a function that counts the number of characters *c* within a string *s* is given. The function scans every position of *s* and updates the counter in case a *c* symbol occurs in the position. The procedure equals the built-in `count` function defined for Python strings.

```
def count_C(s):
    count_c = 0
    for a in s:
        if a == 'c':
            count_c += 1
    return count_c

nof_c = count_C(s)

print(s)
print(nof_c)
print(s.count('c'))
```

```
agctaggaggatcgccagat
4
4
```

Python functions can return multiple values. For example, the following function counts the number of *c* and *g*, and returns the two separated values.

```
def count_CG(s):
    """Count the number of c and g in a given string and return the
counts
    ------
    Parameters:
        s (str) : the input string
    ------
    Returns:
        int : the count of c
        int : the count of g
    """
    count_c, count_g = 0,0
    for a in s:
        if a == 'c':
            count_c += 1
        elif a == 'g':
            count_g += 1
    return count_c, count_g

nof_c, nof_g = count_CG(s)

print(s)
print(nof_c, nof_g)
print(s.count('c'), s.count('g'))
```

```
agctaggaggatcgccagat
4 7
4 7
```

Python functions receive multiple input parameters and return multiple values, and moreover, Python is not a typed programming language, therefore, understanding the correct value to be passed to a function, as well as the meaning and the type of the returned values may result difficult for the final developer who wants to use the function. Thus, documentation of the function plays a crucial role in the usability of Python. The code shown below gives an example of how to document input parameters and return values of a function. Return values do not have names, but the order in which they are returned is crucial for correct comprehension.

```
help(count_CG)
```

```
Help on function count_CG in module __main__:

count_CG(s)
    Count the number of c and g in a given string and return the
counts
    ------
    Parameters:
        s (str) : the input string
    ------
    Returns:
        int : the count of c
        int : the count of g
```

The CG content of a string corresponds to the sum of the number of c and g characters divided by the total length of the string. The following Python function computes the CG content.

```
print("the CG content of "+s+ is "+str((nof_c+nof_g)/len(s)))
print(str((nof_c + nof_g))+" of " +str(len(s)) +" positions")
```

```
the CG content of agctaggaggatcgccagat is 0.55
11 of 20 positions
```

8.4 Counting Words

The previous example showed how to count for specific single characters, however, counting the occurrences of longer words (k-mers) is often required. The `count_occurrences` function counts the number of occurrences of a word w in a string s. For each position of s it verifies if an occurrence of w starts from that position. The verification is made by comparing one by one the characters of w with those occurring in the sibling positions of s.

The built-in python library already provides a string function to count the occurrences of a substring. However, this exercise wants to emphasize the complexity of such a searching approach. It requires two nested `for` loops, which result in a time that is $|s|$ times $|w|$.

More generally, the reader is encouraged to understand the complexity of the built-in functions when they are used, because they can conceal an unexpected complexity in time or space. The function given in the next example is the most trivial way to count substring occurrences, but more powerful methods exist, available in the Python library.

```
def count_occurrences(s,w):
    """
        Count the number of occurrences of w in s
        ---------
        Parameters:
            s (str)
            w (str)
    """
    count = 0
    for i in range(len(s) -len(w) +1):
        for j in range(len(w)):
            if s[i+j] != w[j]:
                break
        else:
            count += 1
        #if s[i:i+len(w)] == w:
        #    count += 1
    return count
```

```
print(count_occurrences(s,'ga'))
print(s.count('ga'))
```

```
3
3
```

When a function to count the number of occurrences of a single word has been defined, the multiplicity of all k-mers occurring in a string can be obtained. The following example uses the multiplicity information to inform about the repetitiveness of all 2-mers (and more in general k-mers) of the string s. The k-mers are distinguished into hapaxes and repeats, and special repeats are duplexes that occur exactly twice.

```
k = 2
kmers = get_kmers(s,k)
for kmer in kmers:
    m = count_occurrences(s,kmer)
    if m== 1:
        print(kmer + " is an hapax")
    else:
        print(kmer + " is a repeat")
        if m== 2:
            print(kmer + " is a duplex")
```

```
cc is an hapax
gc is a repeat
gc is a duplex
ag is a repeat
at is a repeat
at is a duplex
ca is an hapax
tc is an hapax
cg is an hapax
ga is a repeat
gg is a repeat
gg is a duplex
ta is an hapax
ct is an hapax
```

The procedure is modified such that the multiplicity information is not just printed, but words are bucketed depending on their occurrence property. The operation is performed by representing buckets as sets.

```
k = 2
kmers = get_kmers(s,k)

hapaxes = set()
repeats= set()
duplexes = set()

for kmer in kmers:
    m = count_occurrences(s,kmer)
    if m== 1:
```

```
        hapaxes.add(kmer)
    else:
        repeats.add(kmer)
        if m == 2:
            duplexes.add(kmer)

print('hapaxes', len(hapaxes), sorted(hapaxes))
print('repeats', len(repeats), sorted(repeats))
print('duplexes', len(duplexes), sorted(duplexes))
```

```
hapaxes 6 ['ca', 'cc', 'cg', 'ct', 'ta', 'tc']
repeats 5 ['ag', 'at', 'ga', 'gc', 'gg']
duplexes 3 ['at', 'gc', 'gg']
```

By definition, the set of duplexes is contained within the complete set of repeats. A way to retrieve the set of repeats that occur more than twice is to use the built-in python set-theoretic operations, in this case, the set difference is denoted by the symbol −.

```
nodupl = repeats - duplexes
print("repeats not duplexes", len(nodupl), sorted(nodupl))
```

```
repeats not duplexes 2 ['ag', 'ga']
```

Unfortunatelly, the built-in method `count` does not count overlapping occurrences.

```
s = 'aaa'
count_occurrences(s,'aa')
s.count('aa')
```

```
1
```

In fact, the expected number of `aa` in the string `aaa` is three, however, the built-in function only takes into account the first occurrence of the pattern and ignores the overlapped one.

8.5 Searching for Nullomers

Nullomers are k-mers over a specific alphabet that do not appear in a given string. Thus, to search for nullomer, firstly a way to generate the complete set of words is necessary. Then, the presence of the k-mers can be verified.

The complete set can be computed via a recursive function. It extends words of length x to length $x + 1$ in a combinatorial way by elongating them with all the symbols of the alphabet. This means that, given a word α, it generates $\{\alpha a, \alpha c, \alpha g, \alpha t\}$, then it recursively extends such elongations until words of length k are formed. The initial string is set to be the empty string. The stop condition of the recursion is reached when the formed word has a length equal to k.

```
def list_words(prefix, k, alphabet):
    if len(prefix) == k:
        print(prefix)
    else:
        for a in alphabet:
            list_words(prefix + a, k, alphabet)

nuc_alphabet = ['a','c','g','t']

list_words('', 1, nuc_alphabet)
```

```
a
c
g
t
```

```
list_words('', 2, nuc_alphabet)
```

```
aa
ac
ag
at
ca
cc
cg
ct
ga
gc
gg
gt
ta
tc
tg
tt
```

This function can be modified in such a way that k-mers are not printed, but added to an output set.

```
def list_words_2(prefix, k, nuc_alphabet, words):
    if len(prefix) == k:
        words.add(prefix)
    else:
        for a in nuc_alphabet:
            list_words_2(prefix + a, k, nuc_alphabet, words)

kmers = set()
list_words_2('',2, nuc_alphabet, kmers)
print(kmers)
```

```
{'cc', 'gc', 'gt', 'tg', 'ct', 'ag', 'ac', 'at', 'ca', 'cg', 'tc',
'tt', 'ga', 'gg', 'ta', 'aa'}
```

For each k-mer, if its multiplicity within the string is equal to zero it means that it is not occurring in the string, thus it is a nullomer.

```
k = 2
kmers = set()
list_words_2('',k,nuc_alphabet,kmers)

nullomers = set()
for kmer in kmers:
    if count_occurrences(s,kmer) == 0:
        nullomers.add(kmer)

print(s)
print(nullomers)
```

```
agctaggaggatcgccagat
{'ac', 'tg', 'tt', 'gt', 'aa'}
```

The mere presence of a single occurrence can be searched for. If on average k-mer appear more than twice, the just-presence method reduces the time of the procedure since it stops to the first occurrence and does not search for other occurrences.

Python provides a built-in operator, that is inclusion operator `in`, to check for presence of an elements within a set. The operator can also be combined with the logical `not` operator.

```
A = {'a','b','c'}
print('a' in A)
print('a' not in A)
```

```
True
False
```

When the in operator is applied to strings, it searches whenever a string is contained in another.

```
k = 2
kmers = set()
list_words_2('',k,nuc_alphabet,kmers)

nullomers = set()
for kmer in kmers:
    if kmer not in s:
        nullomers.add(kmer)

print(s)
print('nullomers:', nullomers)
```

```
agctaggaggatcgccagat
nullomers: {'ac', 'tg', 'tt', 'gt', 'aa'}
```

8.6 Dictionary Coverage

The **sequence coverage** of a dictionary D over a genome s is defined as the number of positions of the genome that are involved in at least one occurrence of a word of the dictionary.

For each word in the dictionary, we first need to find its occurrences within the genome, and then we need to keep track of the positions that they cover.

The function `get_positions` returns the list of starting positions of the occurrences of a word w in a string s.

An array of boolean values is used to keep track of the coverage of the occurrence. Initially, the array is filled with ‘False’ values, subsequently, it is filled with ‘True’ values for each position covered by the word of the given dictionary.

A dictionary may contain words of different lengths, but this property does not influence the definition of sequence coverage.

```
def get_positions(s,w):
    “ “ “
        Return the starting positions in a reference string s where
        a word w occurs
        ——
        Parameters:
            s (str) : the reference string
            w (str) : the searched word
        ——
        Returns:
            list[int] : the positions
    ” ” ”
    positions = list()
    for i in range(len(s)):
        if s[i:i+len(w)] == w:
            positions.append(i)
    return positions

# s = ‘agctaggaggatcgccagat’
dictionary = [‘ga’, ‘ag’, ‘ca’, ‘t’, ‘aaaa’]

coverage = [False for _ in range(len(s))]

# coverage = [False] * len(s)

for w in dictionary:
    for pos in get_positions(s,w):
        for i in range(len(w)):
            coverage[pos + i] = True

print(s)
print(coverage)

print(‘sequence coverage:’, coverage.count(True),‘covered positions
 over a total of’, len(s),‘=’, coverage.count(True)/len(s))
```

```
agctaggaggatcgccagat
[True, True, False, True, True, True, True, True, True, True, True,
True, False, False, False, True, True, True, True, True]
sequence coverage: 16 covered positions over a total of 20 = 0.8
```

When a k-mer occurs in a given genome, it **covers** k positions, the position of its first character, which is considered the position of its occurrence, and the $(k - 1)$ following positions of the k-mer. The **position coverage** of a given genome, with respect to a given dictionary, is defined as the number of words of the dictionary that cover a given position of the genome.

The computation of the position coverage is obtained by modification of the implementation of sequence coverage. The trace array is converted from a boolean vector to an array of integers. The array is initially filled with zero values. For each word occurrence, the corresponding positions in the array are increased by a value equal to 1.

```
def get_positions(s,w):
    """
        Return the starting positions in a reference string s where
        a word w occurs
        ----
        Paramters:
            s (str) : the reference string
            w (str) : the searched word
        ----
        Returns:
            list[int] : the positions
    """
    positions = list()
    for i in range(len(s)):
        if s[i:i+len(w)] == w:
            positions.append(i)
    return positions

print(s)

dictionary = ['ga', 'ag', 'ca', 't']

coverage = [0 for _ in range(len(s)) ]
for w in dictionary:
    for pos in get_positions(s,w):
        for i in range(len(w)):
            coverage[pos + i] += 1

print(coverage)
```

```
agctaggaggatcgccagat
[1, 1, 0, 1, 1, 1, 1, 2, 1, 1, 1, 1, 0, 0, 0, 1, 2, 2, 1, 1]
```

The sequence coverage can also be retrieved from the integer array by counting the number of cells with a value that differs from zero.

```
print('sequence coverage',(len(coverage) - coverage.count(0)),'/',
        len(coverage),'=',
        (len(coverage) - coverage.count(0)) / len(coverage))
```

```
sequence coverage 16 / 20 = 0.8
```

Statistics concerning the positional coverage for all the positions of the genome can be computed by the following script.

```
from statistics import mean, stdev

print('average positional coverage', mean(coverage))
print('standard deviation of positional coverage', stdev(coverage))
```

```
average positional coverage 0.95
standard deviation of positional coverage 0.6048053188292994
```

List comprehension can be used to calculate the mean positional coverage of only the covered positions.

```
print('average positional coverage of covered positions',
    mean([ i for i in coverage if i > 0]))
print('standard deviation of positional coverage of covered
positions', stdev([ i for i in coverage if i > 0]))
```

```
average positional coverage of covered positions 1.1875
standard deviation of positional coverage of covered positions
    0.4031128874149275
```

8.7 Reading FASTA Files

FASTA files are text files containing genomic sequences written in the FASTA format. Multiple sequences can be stored in a single file. Each sequence is preceded by a description line that starts with a > character. Sequences are split into multiple lines, usually having at most 80 characters per each.

In python, file pointers are provided by the built-in open function (in reading or writing mode). This function returns an object that technically is a pointer to the file but it can also be used as an iterable object. The iteration is performed along the lines of the file.

```
ifile = 'mycoplasma_genitalium_G37.fna'

s = ''

for line in open(ifile, 'r'):
    if line.strip()[0] != '>':
        s += line.strip()

print( len(s) )
```

```
580076
```

The split function removes multiple blank characters (spaces, tabulations and any other printable symbol that results in a blank print) from the left and right of the string.

The first use of strip ensures that the > symbol is checked even if spaces are inserted before. The second one is used to remove the new line character, namely \n from the read line. If the strip is not performed, the resultant genomic sequence will be a string of lines concatenated by \n character. This means that the alphabet of the string will contain also the \n character.

8.8 Informational Indexes

The maximum repeat length (mrl) is defined as the length of the longest repeat. Starting from $k > 1$, the search for the mrl checks for the existence of a repeat of length k, and in a positive case it scans forward to $k + 1$. If no repeat is found at length $k + 1$, then the procedure stops and the returned mrl value is k.

```
def mrl(s):
    """ Calculate the maximal repeat length of a string s """
    k = 0
    mrl = 0
    next_k = True
    while next_k:
        k += 1
        next_k = False
        for kmer in get_kmers(s,k):
            mult = count_occurrences(s,kmer)
            if mult > 1:
                mrl = k
                next_k = True
    return mrl

mrl_s = mrl(s[:1000])
# redefining mrl as a variable will overwrite the mrl function
# definition, thus the name of the variable is mrl_s
print('mrl', mrl_s)
```

```
mrl 14
```

A modified version of the function returns the mrl value together with one of the repeats having such a length. The multiplicity value is also returned.

```
def mrl(s):
    """ Calculate the maximal repeat length of a string s """
    k = 0
    mrl = 0
    kmer_mrl = ''
    mult_mrl = 0
```

```
    next_k = True
    while next_k:
        #print(k, end='', sep=' ')
        k += 1
        next_k = False
        for kmer in get_kmers(s,k):
            mult = count_occurrences(s,kmer)
            if mult > 1:
                mrl = k
                kmer_mrl = kmer
                mult_mrl = mult
                next_k = True
    return mrl, kmer_mrl, mult_mrl


mrl_s, kmer_mrl, mult_mrl = mrl(s[:1000])
# redefining mrl as a variable will overwrite the mrl function
# definition, thus the name of the variable is mrl_s


print('mrl', mrl_s, ', kmer', kmer_mrl, ', multiplicity', mult_mrl)
```

```
mrl 14 , kmer TAACAATATTATTA , multiplicity 2
```

The search for the minimum hapax length (mhl) is similar to the search for mrl. Starting from $k = 1$, the multiplicity of k-mers is examined and the value of k is increased until the first hapax has not been found.

```
def mhl(s):
    """ Calculate the minimal hapax length of a string s """
    k = 0
    mhl = 0
    kmer_mhl = ''
    mult_mhl = 0
    next_k = True
    while next_k:
        k += 1
        for kmer in get_kmers(s,k):
            mult = count_occurrences(s,kmer)
            if mult == 1:
                mhl = k
                kmer_mhl = kmer
                mult_mhl = mult
                next_k = False
    return mhl, kmer_mhl, mult_mhl


mhl_s, kmer_mhl, mult_mhl = mhl(s[:1000])

print('mhl', mhl_s, ', kmer', kmer_mhl, ', multiplicity', mult_mhl)
```

```
mhl 3 , kmer CGA , multiplicity 1
```

The minimum forbidden length is calculated by comparing the size of the effective set of k-mers occurring in a sequence with 4^k. The smallest value of k at which the two sizes differ is the wanted value.

```
def get_alphabet(s):
    al = set()
    for c in s:
        al.add(c)
    return al

def mfl(s, alphabet = None):
    """ Calculate the minimal forbidden length of a string s """
    if alphabet == None:
        a = len(get_alphabet(s))
    else:
        a = len(alphabet)

    k = 0
    while True:
        k += 1
        kmers = get_kmers(s,k)
        if len(kmers) != a**k:
            return k


mfl_s = mfl(s)


print('mfl',mfl_s, ', mcl', mfl_s - 1)
```

```
mfl 6 , mcl 5
```

The reader may notice that in the previous examples, the variable used to store the calculated indexes never has the same name as the function used to calculate it. This is due to the fact that if a variable with the same name is used, this fact overwrites the declaration of the function, therefore calling the function produces an error exception, as it is shown in the following example.

```
def my_funct():
    return 0

my_funct = my_funct()
# declaring a variable having the same name of a function
# declared before
print(my_funct)
```

```
0
```

```
my_funct()
```

```
---------------------------------------------------------------------------
TypeError                                 Traceback (most recent call last)
<ipython-input-13-50af49a3d99d> in <module>()
```

```
my_funct()
TypeError: 'int' object is not callable
```

Another useful informational index is also the empirical k-entropy. In order to compute (empirical) k-entropy, multiplicities of words are converted into probabilities (frequencies) and then classical Shannon's formula is used:

```
import math

def k_entropy(s,k):
    """ Calculate the empirical entropy at word length k of a
        string s """

    t = 0.0
    for kmer in get_kmers(s,k):
        t += count_occurrences(s,kmer)

    e = 0.0
    for kmer in get_kmers(s,k):
        e += math.log(count_occurrences(s,kmer) / t, 2)
    return -e

k = 2
print(str(k)+'-entropy', k_entropy(s,k))
```

```
2-entropy 68.1465951013701
```

8.9 Genomic Distributions

In this section, Python programs are presented that extract the most important distributions of informational genomics.

Word Multiplicity Distribution (WMD)

The word multiplicity distribution (WMD) is a function that assigns to any word its corresponding multiplicity in a given string. In the following example, to each symbol of the alphabet, the number of its occurrences within a string is assigned.

```
nuc_alphabet = ['a','c','g','t']
for n in nuc_alphabet:
    print(n, count_occurrences(s,n))
```

```
a 6
c 4
g 7
t 3
```

Multiplicities for 2-mers occurring in a string $s =' agctaggaggatcgccagat'$ are given below.

```
kmers = get_kmers(s, k = 2)
for kmer in kmers:
    print(kmer, count_occurrences(s,kmer))
```

```
tc 1
cg 1
ca 1
ta 1
ct 1
gc 2
ga 3
gg 2
at 2
cc 1
ag 4
```

WMD as key-value collection

The word multiplicity distribution provides a python dictionary which assigns multiplicity values (integers) to a set of key k-mers (strings).

```
WMD= dict() # multiplicity distribution
kmers = get_kmers(s,2)
for kmer in kmers:
    WMD[kmer]  = count_occurrences(s,kmer)

for kmer in WMD.keys():
    print(kmer, WMD[kmer])
```

```
tc 1
cg 1
ga 3
gg 2
ta 1
ct 1
at 2
ca 1
cc 1
ag 4
gc 2
```

There is no assumption about the order in which keys, or pairs, are iterated because it depends on the internal data structure used by the specific python version to implement the dictionary. Thus, in case the user wants to output the distribution in the lexicographical order of the k-mers, explicit sorting must be performed.

```
for kmer in sorted(WMD.keys()):
    print(kmer, WMD[kmer])
```

```
ag 4
at 2
ca 1
cc 1
cg 1
ct 1
ga 3
gc 2
gg 2
ta 1
tc 1
```

Alternatively, the `dict`'s iterator returns a list of pairs `(key,value)` that can be automatically decomposed into the variables `kmer` and `mult`.

```
for kmer,mult in sorted(WMD.items()):
    print(kmer, mult)
```

```
ag 4
at 2
ca 1
cc 1
cg 1
ct 1
ga 3
gc 2
gg 2
ta 1
tc 1
```

Retrieving the set of k-mers and then searching for their multiplicity results to be a redundant combination of steps. Namely, the distribution can be built while scanning for k-mers in the string and, at the same time, by updating their multiplicity values.

```
WMD = dict()
k = 2
for i in range(len(s) -k +1):
        # extraction of the k-mer at position i
        w = s[i:i+k]
        # update of the occurrence of the k-mer w
        WMD[w] += 1
```

```
---------------------------------------------------------------------------
KeyError                                  Traceback (most recent call last)
<ipython-input-38-8cbf6223371f> in <module>()
      3 for i in range(len(s) -k +1):
      4     # extraction of the k-mer at position i
      5     w = s[i:i+k]
      6     # update of the occurrence of the k-mer w
----> 7     WMD[w] += 1

KeyError: 'ag'
```

The single-slice operator of `dict` can only be used to access pairs that have been previously inserted in the collection. Thus, using it to access a key not yet present will cause an error.

A possible way to solve the problem is to first check for the presence of the key, and in case of absence make a first insertion of it. The operator 'in' allows one to check for the existence of a key in a dictionary.

```
WMD= dict()
k = 2
for i in range(len(s) -k +1):
        w = s[i:i+k]
        if w in WMD:
            WMD[w] += 1
        else:
            WMD[w] = 1

for kmer in WMD.keys():
    print(kmer, WMD[kmer])
```

```
tc 1
cg 1
ga 3
gg 2
ta 1
ag 4
at 2
ca 1
cc 1
gc 2
ct 1
```

Alternatively, the built-in `get` function can be used, which can use a default value that is returned in case of a missing key, without causing any error.

In the next example, the default value of zero is used. If *w* has not yet been inserted as a key, the `get` operation returns 0 and the final counter is set to 1. Conversely, if *w* already exists within the collection, the operation returns its current counter that will be updated by 1.

```
WMD= dict()
k = 2
for i in range(len(s) -k +1):
        w = s[i:i+k]
        WMD[w] =WMD.get(w, 0) + 1

for kmer in WMD.keys():
    print(kmer, WMD[kmer])
```

```
tc 1
cg 1
ga 3
gg 2
ta 1
```

```
ag 4
at 2
ca 1
cc 1
gc 2
ct 1
```

Once a working code has been written, it can be also encapsulated in a function.

```
def get_multiplicity_distribution(s,k):
    """
    
        Return the word multiplciity distribution of k-mers
        occurring in the string s
        -------
        Parameters:
            s (str) : the input string
            k (int) : the length of the k-mers
        --------
        Returns:
            dict[str,int] : a dictionary which associates
            multiplicity values to the k-mers in s
    """
    WMD = dict()
    for i in range(len(s) -k +1):
        w = s[i:i+k]
        WMD[w] = WMD.get(w,0) + 1
    return WMD

WMD = get_multiplicity_distribution(s,2)

for k,v in sorted(WMD.items()):
    print(k,v)
```

```
ag 4
at 2
ca 1
cc 1
cg 1
ct 1
ga 3
gc 2
gg 2
ta 1
tc 1
```

The given WMD implementation speeds the calculation of some of the informational indexes up. For example, in the calculation of k-entropies, which in the following example scales down from more than 9 seconds to less than one second.

```
import math

def k_entropy(s,k):
    """
        Calculate the empirical k-entrpy on k-mers in s
```

```
    """
    t = 0.0
    for kmer in get_kmers(s,k):
        t += count_occurrences(s,kmer)

    e = 0.0
    for kmer in get_kmers(s,k):
        e += math.log(count_occurrences(s,kmer) / t, 2)
    return -e


def fast_k_entropy(s,k):
    """
        Calculate the empirical k-entrpy on k-mers in s
    """
    distr = get_multiplicity_distribution(s,k)
    t = sum(distr.values())
    e = 0.0
    for v in distr.values():
        e += math.log(v / t, 2)
    return -e

k = 2

import time

start_time = time.time()
print(str(k)+'-entropy', k_entropy(mg37,k))
print('seconds:', time.time() - start_time)

print()

start_time = time.time()
print(str(k)+'-entropy', fast_k_entropy(mg37,k))
print('seconds:', time.time() - start_time)
```

```
2-entropy 68.14659510137011
seconds: 9.379900455474854

2-entropy 68.14659510137011
seconds: 0.2494194507598877
```

Word length distribution

The word length distribution (WLD) assigns to each word length k the number of k-mers of G, namely the size of $D_k(G)$. It can be calculated by a function that, for a range of word lengths, uses the predeclared `get_kmers` procedure. The following source code shows also how to graphically visualize the distribution as a histogram by means of the `matplotlib` library. The histogram reports for each word length k, on the x-axis, the cardinality of the corresponding dictionary in the analysed sequence.

```
def wld(s, k_start, k_end):
    """

        Calculate the word length distribution of the string s for
        the given range of values of word length k
        ----
        Parameters:
            s (str) : the input string
            k_start (int) : the initial word length
            k_end (int) : the final word length
        ----
        Returns:
            dict[int,int] : a dictionary which associates world
            lengths (key) to the number of k-mers at each length
            (value)
    """

    wld = dict()
    for k in range(k_start,k_end):
        wld[k] = len(get_kmers(s,k))
    return wld

k_start = 1
k_end = 20
wld = wld(mg37, k_start, k_end)

import matplotlib.pyplot as plt
bar_values = [v for k,v in sorted(wld.items())]
plt.rcParams['figure.figsize'] = [20, 6]
plt.bar(range(k_start,k_end), bar_values)
plt.xticks(range(k_start,k_end), range(k_start,k_end))
plt.ylabel('|D_k(G)|')
plt.xlabel('k')
plt.title('Word length distribution')
plt.show()
```

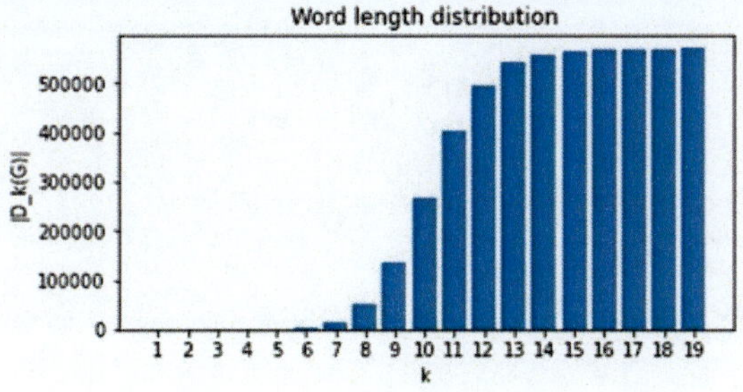

Average multiplicity distribution

The average multiplicity distribution (AMD) assigns to each word length k the average multiplicity of the k-mers of G.

```
import statistics

def amd(s, k_start, k_end):
    """

        Calculate the average multiplicity distribution of the
        string s for the given range of values of word length k
        ----
        Paramters:
            s (str) : the input string
            k_start (int) : the initial word length
            k_end (int) : the final word length
        ----
        Returns:
            dict[int,int] : a dictionary which associates word
            lengths (key) to the average multiplicity at the
            specific word length (value)
    """
    amd = dict()
    for k in range(k_start,k_end):
        amd[k] = statistics.mean(
            get_multiplicity_distribution(s,k).values() )
    return amd

k_start = 1
k_end = 20
amd = amd(mg37, k_start, k_end)

#for k,v in amd.items():
#     amd[k] = math.log(v)

#k_start = 12

import matplotlib.pyplot as plt
bar_values = [v for k,v in sorted(amd.items())]
plt.rcParams['figure.figsize'] = [20, 6]
plt.bar(range(k_start,k_end), bar_values)
plt.xticks(range(k_start,k_end), range(k_start,k_end))
plt.ylabel('Average multiplicity')
plt.xlabel('k')
plt.title('Average multiplicity distribution')
plt.show()
```

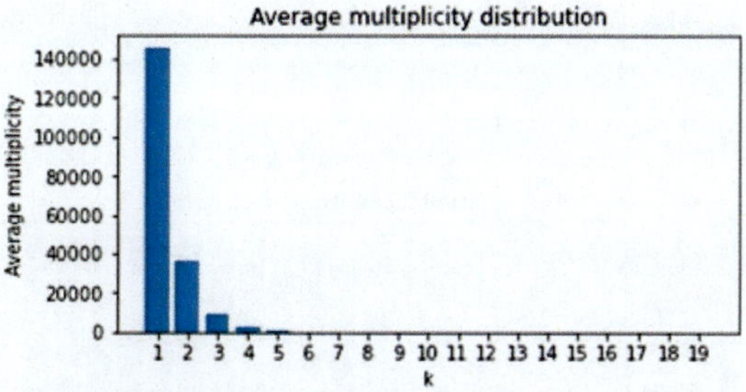

Empirical entropy distribution

The empirical entropy distribution (EED) assigns to each word length *k* the value of the k-entropy.

```
def eed(s, k_start, k_end):
    """
        Calculate the empirical entropy distribution of the string
        s for the given range of values of word length k
        ----
        Paramters:
            s (str) : the input string
            k_start (int) : the initial word length
            k_end (int) : the final word length
        ----
        Returns:
            dict[int,int] : a dictionary which associates word
            lengths (key) to the empirical entropy at the specific
            word length (value)
    """
    eed = dict()
    for k in range(k_start,k_end):
        eed[k] = fast_k_entropy(s,k)
    return eed

k_start = 1
k_end = 20
eed = eed(mg37[:10000], k_start, k_end)

#for k,v in amd.items():
#    amd[k] = math.log(v)

#k_start = 12

import matplotlib.pyplot as plt
bar_values = [v for k,v in sorted(eed.items())]
plt.rcParams['figure.figsize'] = [20, 6]
plt.bar(range(k_start,k_end), bar_values)
plt.xticks(range(k_start,k_end), range(k_start,k_end))
```

```
plt.ylabel('k–entropy')
plt.xlabel('k')
plt.title('Empirical entropy distribution')
plt.show()
```

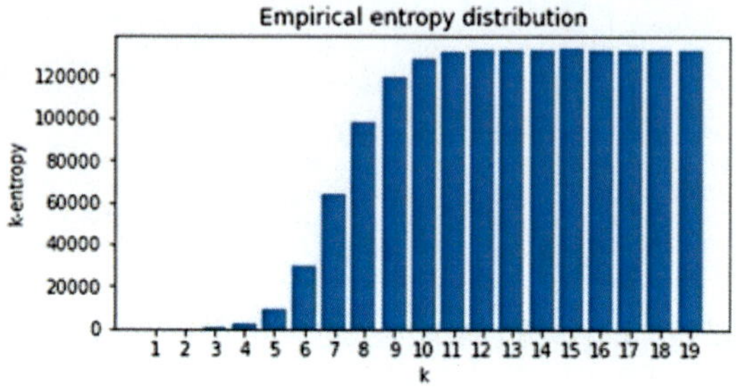

Word co-multiplicity distribution

Given a value of k, the word co-multiplicity distribution (WCMD) reports for each value of multiplicity the number of words having such multiplicity. In the following example, it is calculated for the first 100,000 bases of the *m. genitalium* G37 genome. The source code exploits the predefined `get_multiplicity_distribution` function in order to extract the required information. In fact, given the multiplicity distribution, which assigns to each word presented in the genome its multiplicity, the co-multiplicity distribution is calculated by counting the number of times a given multiplicity appears in the multiplicity distribution.

```
def wcmd(s,k):
    """

        Calculate the word co–multiplicity distribution of the
        string s for the given value of word length k
        ––––
        Paramters:
            s (str) : the input string
            k (int) : the word length
        ––––
        Returns:
            dict[int,int] : a dictionary which associates a
            multiplicity value (key) to the number of k–mers
            having such multiplicity (value)
    """

    distr = dict()
    mdistr = get_multiplicity_distribution(s,k)
    for m in mdistr.values():
        distr[m]= distr.get(m,0) + 1
    return distr

k = 6
wcmd = wcmd(mg37[:100000],k)
```

```
# add missing multiplicity values
for i in range(1,max(wcmd.keys())):
    wcmd[i] = wcmd.get(i,0) + 0

import matplotlib.pyplot as plt
bar_values = [v for k,v in sorted(wcmd.items())]
plt.rcParams['figure.figsize'] = [20, 6]
plt.bar(sorted(wcmd.keys()), bar_values, width=1.0)
plt.ylabel('Number of words')
plt.xlabel('Multiplicity')
plt.title('Word co-multiplicity distribution')
plt.show()
```

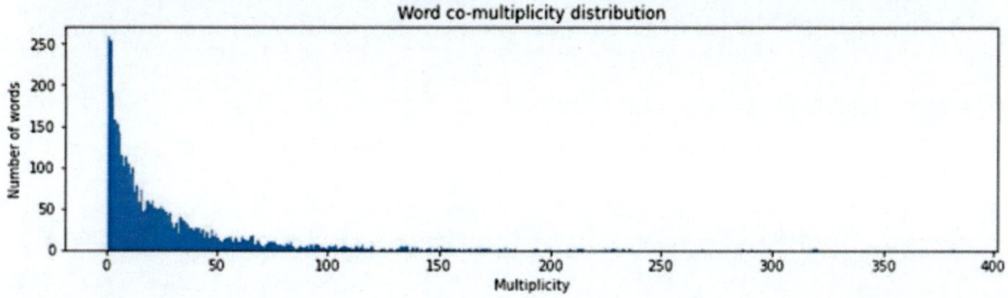

8.10 Genomic Data Structures

The computation of some informational indexes would be prohibitive for real long genomes, by using ingenious algorithms, while if suitable data structures are employed, computations become possible. One of the most important data structures, in the field of informational genomics, is the suffix array, which we present in the following subsection. Additional data structures, such as the longest common prefix (LCP) array, are added up to the suffix array in order to allow the extraction of genomic dictionaries in linear time. Moreover, it has to be pointed out that several state-of-the-art algorithms are able to compute both the suffix array and the LCP array in linear time, however, since they are not the focus of this book, the user is referred to external readings.

Suffix array

The suffix array is the most-used data structure to index texts (strings over any alphabet) such that operations like the search of a substring are made more efficiently with respect to online approaches that do not use indexing structures.

Given a string s and the set of all the suffixes of s, the suffix array reports the starting position of the ith suffix in the lexicographical order.

The lexicographic order over strings has been defined in the Chap. 4, Informational Genomics. However, many variants are used in mathematics, linguistics, and computer science. The standard version, we call **strong lexicographic order**, gives a preliminary order condition where any string precedes the longer ones, while strings of the same length are compared according to the first position, from the left, where they disagree, and the order of their symbols in that position gives the order of strings. A second variant, also called **alphabetic order**, is the lexicographic order used in the context of suffix arrays. In this version, the preliminary order by length is avoided, and strings are directly compared according to their first disagreement position. However, in this case, infinite chains between two strings are possible, for example, if $0 < 1$, for any natural number n strings 0^n precede 1.

Given a string s, the set of its suffixes can be extracted in the following manner:

```
s = 'agctagctagctagtttagct'

suffixes = list()

for i in range(len(s)):
    suffixes.append(s[i:])

for suff in suffixes:
    print(suff)
```

```
agctagctagctagtttagct
gctagctagctagtttagct
ctagctagctagtttagct
tagctagctagtttagct
agctagctagtttagct
gctagctagtttagct
ctagctagtttagct
tagctagtttagct
agctagtttagct
gctagtttagct
ctagtttagct
tagtttagct
agtttagct
gtttagct
tttagct
ttagct
tagct
agct
gct
ct
t
```

Then, the suffixes can be lexicographically sorted via the python function `sorted`.

```
for suff in sorted(suffixes):
    print(suff)
```

```
agct
```

```
agctagctagctagtttagct
agctagctagtttagct
agctagtttagct
agtttagct
ct
ctagctagctagtttagct
ctagctagtttagct
ctagtttagct
gct
gctagctagctagtttagct
gctagctagtttagct
gctagtttagct
gtttagct
t
tagct
tagctagctagtttagct
tagctagtttagct
tagtttagct
ttagct
tttagct
```

NOTE: the strong lexicographical order can be obtained by using the parameter `key` of the function `sorted`. In this manner, the words are sorted by following the order driven by the function specified in the parameter, and in case of equality the lexicographical order is used.

```python
qllist = ['a','b','ab','ac','bb','abc', 'bba','bbc']

for suff in sorted(qllist, key = len):
    print(suff)
```

```
a
b
ab
ac
bb
abc
bba
bbc
```

A trivial and inefficient way to construct the suffix array is to build a list of pairs, where the first element of the pair is one of the suffixes and the second element is its starting position. Because starting positions are all different, the `sorted` function sorts the pairs by the lexicographical order of the suffixes.

```python
s = 'agctagctagctagtttagct'

pairs = list()

for i in range(len(s)):
    pairs.append( (s[i:] ,i) )

for p in sorted(pairs):
    print(p)
```

```
('agct', 17)
('agctagctagctagtttagct', 0)
('agctagctagtttagct', 4)
('agctagtttagct', 8)
('agtttagct', 12)
('ct', 19)
('ctagctagctagtttagct', 2)
('ctagctagtttagct', 6)
('ctagtttagct', 10)
('gct', 18)
('gctagctagctagtttagct', 1)
('gctagctagtttagct', 5)
('gctagtttagct', 9)
('gtttagct', 13)
('t', 20)
('tagct', 16)
('tagctagctagtttagct', 3)
('tagctagtttagct', 7)
('tagtttagct', 11)
('ttagct', 15)
('tttagct', 14)
```

Once the pairs are sorted, the suffix array is built by iterating the sorted pairs and by extracting their starting position (namely the second element of the pair).

```
print('suffixes'+ ' '*(len(s) - len('suffixes')) + 'suffix_array')

for p in sorted(pairs):
    print(p[0] + ' '*(p[1]) ,p[1])
    # a given number of space characters (' '*(p[1])) is added
    # such that the elements of the suffix array are print on the
    # same column
```

```
suffixes              suffix_array
agct                  17
agctagctagctagtttagct 0
agctagctagtttagct     4
agctagtttagct         8
agtttagct             12
ct                    19
ctagctagctagtttagct   2
ctagctagtttagct       6
ctagtttagct           10
gct                   18
gctagctagctagtttagct  1
gctagctagtttagct      5
gctagtttagct          9
gtttagct              13
t                     20
tagct                 16
tagctagctagtttagct    3
tagctagtttagct        7
tagtttagct            11
ttagct                15
tttagct               14
```

Actually, we are not interested in storing the ordered set of suffixes, but only in storing the resultant suffix array. Thus, the function `get_suffix_array` receives in input a string s and returns only the array.

```python
def get_suffix_array(s):
    """
    Construct the suffix array of the string s.
    """
    pairs = list()
    for i in range(len(s)):
        pairs.append( (s[i:],i) )
    sa = list()
    for p in sorted(pairs):
        sa.append(p[1])
    return sa

sa = get_suffix_array(s)
```

The list of sorted suffixes can be obtained by iterating over the suffix array and by extracting on the fly the suffix corresponding to the given starting position given by the array. Let sa be the suffix array of a string s, then the ith ordered suffix starts in position $sa[i]$, and the suffix corresponds to $s[i:]$. A padding of extra space characters `' '*(sa[i])` is added to the suffix for a well-print format.

```python
def print_suffix_array(s, sa):
    for i in range(len(sa)):
        print(s[sa[i]:] + ' '*(sa[i]), sa[i])

print_suffix_array(s,sa)
```

```
agct                  17
agctagctagctagtttagct 0
agctagctagtttagct     4
agctagtttagct         8
agtttagct             12
ct                    19
ctagctagctagtttagct   2
ctagctagtttagct       6
ctagtttagct           10
gct                   18
gctagctagctagtttagct  1
gctagctagtttagct      5
gctagtttagct          9
gtttagct              13
t                     20
tagct                 16
tagctagctagtttagct    3
tagctagtttagct        7
tagtttagct            11
ttagct                15
tttagct               14
```

LCP intervals and enhanced suffix arrays (ESA)

Indexing structures based on suffix arrays are enhanced by means of a further array reporting the length of the Longest Common Prefix (LCP) between two consecutive suffixes in the lexicographical order. The LCP value of the first suffix is set to be 0. The length of the common prefix between the suffixes can be computed in a trivial way as it is shown in the function `longest_prefix_length`. The function computes the length for the two suffixes starting at position i and j of the input string s. On top of this function, the `get_lcp` procedure is defined.

```
def longest_prefix_length(s, i, j):
    """
        Calculate the length of the longest common prefix between
        two suffixes, the one in position i and the other in
        position j, of s
    """
    l = 0
    while (i+l < len(s)) and (j+l < len(s)):
        if s[i+l] != s[j+l]:
            break
        l += 1
    return l


def get_lcp(s,sa):
    """
        Construct the LCP array associated to the suffix array
        (sa)of the string s.
        The LCP value of the first suffix is set to be 0.
    """
    lcp = list()
    lcp.append(0)
    for i in range(1,len(sa)):
        lcp.append( longest_prefix_length(s, sa[i], sa[i-1]) )
    return lcp


lcp = get_lcp(s,sa)
```

The print of the ordered suffixes together with the suffix array is extended to report the lcp values.

```
def print_sa_lcp(s,sa,lcp):
    print('index', 'suffixes' + ' '*(len(s)-len('suffixes')), 'SA',
     'LCP', sep='\t')
    print('-'*45)
    for i in range(len(sa)):
        print(i, s[sa[i]:] + ' '*(sa[i]), sa[i], lcp[i], sep='\t')

print_sa_lcp(s,sa,lcp)
```

```
index   suffixes                SA  LCP
----------------------------------------
0    agct                       17  0
1    agctagctagctagtttagct      0   4
2    agctagctagtttagct          4   10
3    agctagtttagct              8   6
4    agtttagct                  12  2
5    ct                         19  0
6    ctagctagctagtttagct        2   2
7    ctagctagtttagct            6   8
8    ctagtttagct                10  4
9    gct                        18  0
10   gctagctagctagtttagct       1   3
11   gctagctagtttagct           5   9
12   gctagtttagct               9   5
13   gtttagct                   13  1
14   t                          20  0
15   tagct                      16  1
16   tagctagctagtttagct         3   5
17   tagctagtttagct             7   7
18   tagtttagct                 11  3
19   ttagct                     15  1
20   tttagct                    14  2
```

An LCP k-interval is defined as a contiguous region of the LCP array, defined by two indexes i and j (with $1 \leq i \leq j \leq |s|$) such that for $i < x \leq j$, $LCP[x] \geq k$. Moreover, $LCP[i-1] < k$ and $LCP[j+1] < k$.

LCP intervals are useful in enumerating the k-mers occurring in a given string. In fact, for $k = 1$ all and only the suffixes starting with the symbol 'a' are contiguously listed in the first positions of the suffix array. Moreover, such suffixes have a lcp value greater than 1, except for the first one. For the second suffix, the first position where the lcp value is less than 1 is the one identified by the suffix 'ct'. More in general an lcp value of 0 occurs each time the corresponding suffix has the first character different from the previous suffix.

In the previous example, the four nucleobases A, C, G, T are recognized by 4 lcp 1-intervals that are [0, 4], [5, 8], [9, 13] and [14, 20]. Within each interval, the lcp value i is greater or equal to 1, except for the first element that has an lcp value of 0 and that identifies the end of the previous interval and the start of the current one.

An algorithm to retrieve LCP k-intervals uses two indexes, i and j, to identify the bounds of the current interval. The i index is the start, and the j index is the end (not included).

The algorithm starts with $i = 0$ and $j = 1$, increases the value of j until $lcp[j] \geq k$ (and until j is a valid index, namely it is less than the length of the array). When $lcp[j] < k$, the algorithm stops and return the interval $[i, j]$. Subsequently, i is set to be the current value of j, j is increased by 1 and the search is repeated. The algorithm continues until i is less than the length of the array.

```python
def print_sa_lcp_region(s,sa,lcp, i,j):
    print('-'*40)
    print('interval')
```

```
    for x in range(i,j):
        print(x,s[sa[x]:] +' '*(sa[x]), sa[x], lcp[x], sep='\t')
    #print('.'*40)


k = 1
i = 0
while i < len(s):
    j = i + 1
    while (j < len(s)) and (lcp[j] >= k):
        j += 1

    print_sa_lcp_region(s,sa,lcp, i,j)
    print('k-mer:', s[ sa[i]:sa[i]+k] )

    i = j
```

```
________________________________________
interval
0	agct	17	0
1	agctagctagctagtttagct	0	4
2	agctagctagtttagct	4	10
3	agctagtttagct	8	6
4	agtttagct	12	2
k-mer: a
________________________________________
interval
5	ct	19	0
6	ctagctagctagtttagct	2	2
7	ctagctagtttagct	6	8
8	ctagtttagct	10	4
k-mer: c
________________________________________
interval
9	gct	18	0
10	gctagctagctagtttagct	1	3
11	gctagctagtttagct	5	9
12	gctagtttagct	9	5
13	gtttagct	13	1
k-mer: g
________________________________________
interval
14	t	20	0
15	tagct	16	1
16	tagctagctagtttagct	3	5
17	tagctagtttagct	7	7
18	tagtttagct	11	3
19	ttagct	15	1
20	tttagct	14	2
k-mer: t
```

The previous algorithm results to be not suitable for values of k greater than 1, in fact when it is applied to extract 2-mers unwanted intervals, thus unwanted words, are retrieved. In particular, the 1-mer t is produced, which is due to the fact that the algorithm extracts prefixes that are shorter than the given value of k.

```
k = 2
i = 0
while i < len(s):
    j = i + 1
    while (j < len(s)) and (lcp[j] >= k):
        j += 1

    print_sa_lcp_region(s,sa,lcp, i,j)
    print('k-mer:', s[ sa[i]:sa[i]+k] )

    i = j
```

```
____________________________________________
interval
0   agct                    17  0
1   agctagctagctagtttagct   0   4
2   agctagctagtttagct       4   10
3   agctagtttagct           8   6
4   agtttagct               12  2
k-mer: ag
____________________________________________
interval
5   ct                      19  0
6   ctagctagctagtttagct     2   2
7   ctagctagtttagct         6   8
8   ctagtttagct             10  4
k-mer: ct
____________________________________________
interval
9   gct                     18  0
10  gctagctagctagtttagct    1   3
11  gctagctagtttagct        5   9
12  gctagtttagct            9   5
k-mer: gc
____________________________________________
interval
13  gtttagct                13  1
k-mer: gt
____________________________________________
interval
14  t                       20  0
k-mer: t
____________________________________________
interval
15  tagct                   16  1
16  tagctagctagtttagct      3   5
17  tagctagtttagct          7   7
18  tagtttagct              11  3
k-mer: ta
```

```
----------------------------------------
interval
19  ttagct                  15  1
20  tttagct                 14  2
k-mer: tt
```

In order to avoid the extraction of such extra intervals, the algorithm has to take into account that right intervals are not contiguous but are interleaved with intervals that must be discarded. The condition for discarding such intervals is given by the check of their proximity with the end of the string.

```python
k = 2
i = 0
while i < len(s):
    while (i < len(s)) and (sa[i] > len(s) - k - 1): # check
        i += 1
    if i == len(s): # there are no more valid intervals
        break

    j = i + 1
    while (j < len(s)) and (lcp[j] >= k):
        j += 1

    print_sa_lcp_region(s,sa,lcp, i,j)
    print('k-mer:', s[ sa[i]:sa[i]+k] )

    i = j
```

```
----------------------------------------
interval
0   agct                    17  0
1   agctagctagctagtttagct   0   4
2   agctagctagtttagct       4   10
3   agctagtttagct           8   6
4   agtttagct               12  2
k-mer: ag
----------------------------------------
interval
6   ctagctagctagtttagct     2   2
7   ctagctagtttagct         6   8
8   ctagtttagct             10  4
k-mer: ct
----------------------------------------
interval
9   gct                     18  0
10  gctagctagctagtttagct    1   3
11  gctagctagtttagct        5   9
12  gctagtttagct            9   5
k-mer: gc
----------------------------------------
interval
13  gtttagct                13  1
k-mer: gt
```

```
________________________________________
interval
15  tagct                    16  1
16  tagctagctagtttagct        3  5
17  tagctagtttagct            7  7
18  tagtttagct               11  3
k-mer: ta
________________________________________
interval
19  ttagct                   15  1
20  tttagct                  14  2
k-mer: tt
```

Some notes on the enhanced suffix array

1. Due to the lexicographical construction of the suffix array, the enumeration of k-mer via lcp k-intervals implicitly follows the lexicographical order of enumerating the k-mers.
2. The enhanced suffix array implicitly contains the complete set $D(s)$, which includes every k-mer occurring in s for $1 \leq k \leq |s|$.

Informational genomics via ESA (and NESA)

Usually, sequenced genomes present an extra symbol, coded with an N, which represents an ambiguity in determining a specific nucleotide at a given position of the genome. The execution of the previous approach on such sequences produces k-mers that may contain N symbols. However, it is desirable to skip such k-mers from the enumeration.

The issue is solved with an approach similar to the one used to discard k-mers shorter than the desired word length k. A modified algorithm discards suffixes which have a symbol N within the initial k characters.

In order to increase the efficiency in time complexity, a further array is computed. The array, called NS, takes a trace of the distance from the starting position of the suffix to the closest N character on the right of such position.

```
def distance_to_n(s,i):
    j = i
    while (j<len(s)) and (s[j] != 'N'):
        j += 1
    return j - i

def get_ns_array(s,sa):
    return [  distance_to_n(s,sa[i]) for i in range(len(s)) ]

```

```python

def print_sa_lcp_ns(s,sa,lcp, ns):
    print('index', 'suffixes' + ' '*(len(s)-len('suffixes')), 'SA',
     'LCP', 'NS', sep='\t')
    print('-'*60)
    for i in range(len(sa)):
        print(i, s[sa[i]:] + ' '*(sa[i]), sa[i], lcp[i], ns[i],
        sep='\t')



s = 'agctagNctagctagNtttagctN'


sa = get_suffix_array(s)
lcp = get_lcp(s,sa)
ns = get_ns_array(s,sa)


print_sa_lcp_ns(s,sa,lcp,ns)
print('-'*60)
```

```
index   suffixes                        SA  LCP NS
------------------------------------------------------------
0   N                           23  0   0
1   NctagctagNtttagctN          6   1   0
2   NtttagctN                   15  1   0
3   agNctagctagNtttagctN        4   0   2
4   agNtttagctN                 13  3   2
5   agctN                       19  2   4
6   agctagNctagctagNtttagctN    0   4   6
7   agctagNtttagctN             9   7   6
8   ctN                         21  0   2
9   ctagNctagctagNtttagctN      2   2   4
10  ctagNtttagctN               11  5   4
11  ctagctagNtttagctN           7   4   8
12  gNctagctagNtttagctN         5   0   1
13  gNtttagctN                  14  2   1
14  gctN                        20  1   3
15  gctagNctagctagNtttagctN     1   3   5
16  gctagNtttagctN              10  6   5
17  tN                          22  0   1
18  tagNctagctagNtttagctN       3   1   3
19  tagNtttagctN                12  4   3
20  tagctN                      18  3   5
21  tagctagNtttagctN            8   5   7
22  ttagctN                     17  1   6
23  tttagctN                    16  2   7
------------------------------------------------------------
```

Two further conditions are included on enumerating k-mers.

The first one is due to a modification of the definition of the lcp k-interval such that it is extended by the condition $LCP[x] \geq k$. This ensures that interval do not contains k-mers having a N character in the initial k positions.

The second condition is applied to discard unwanted intervals that interleave valid ones.

```
def print_sa_lcp_ns_region(s,sa,lcp,ns, i,j):
    print('-'*60)
    print('interval')
    for x in range(i,j):
        print(x,s[sa[x]:] +' '*(sa[x]), sa[x], lcp[x], ns[x],
        sep='\t')


k = 3
i = 0
while i < len(s):
    while (i < len(s)) and ( (sa[i] > len(s) - k - 1)
            or (ns[i] < k) ): # second further condition
        i += 1
    if i == len(s):
        break

    j = i+1
    while (j < len(s)) and (lcp[j] >= k)
            and (ns[i] >= k): # first further condition
        j += 1

    print_sa_lcp_ns_region(s,sa,lcp,ns, i,j)
    print('k-mer:', s[ sa[i]:sa[i]+k] )
    i = j
```

```
------------------------------------------------------------
interval
5   agctN                       19  2   4
6   agctagNctagctagNtttagctN    0   4   6
7   agctagNtttagctN             9   7   6
k-mer: agc
------------------------------------------------------------
interval
9   ctagNctagctagNtttagctN      2   2   4
10  ctagNtttagctN               11  5   4
11  ctagctagNtttagctN           7   4   8
k-mer: cta
------------------------------------------------------------
interval
14  gctN                        20  1   3
15  gctagNctagctagNtttagctN     1   3   5
16  gctagNtttagctN              10  6   5
k-mer: gct
------------------------------------------------------------
interval
18  tagNctagctagNtttagctN       3   1   3
```

```
19  tagNtttagctN        12  4  3
20  tagctN              18  3  5
21  tagctagNtttagctN     8  5  7
k-mer: tag
________________________________________________
interval
22  ttagctN             17  1  6
k-mer: tta
________________________________________________
interval
23  tttagctN            16  2  7
k-mer: ttt
```

The following code provides a faster implementation of the construction of the N array. It constructs an inverse suffix array, such that for each position i in the string, the corresponding position of the i-th suffix in the suffix array is obtained in constant time. Then, starting from the end of the string, it keeps track of the last position, on the right, where the symbol N has been found, thus it assigns the distance between the current position and the last right occurrence of the N.

```
def fast_get_ns_array(s, sa):
    inv_sa = [0 for _ in range(len(sa))]
    for i in range(len(sa)):
        inv_sa[ sa[i] ] = i

    ns = [0 for _ in range(len(sa))]
    lastn = len(s)
    for i in range(len(s)-1,-1,-1):
        if s[i] == 'N':
            lastn = i
        ns[ inv_sa[i] ] = lastn - i
    return ns


fns = fast_get_ns_array(s, sa)
print(len(fns), fns)

ns = get_ns_array(s, sa)
print(len(ns), ns)

assert ns == fns
```

```
24 [0, 0, 0, 2, 2, 4, 6, 6, 2, 4, 4, 8, 1, 1, 3, 5, 5, 1, 3, 3, 5,
7, 6, 7]
24 [0, 0, 0, 2, 2, 4, 6, 6, 2, 4, 4, 8, 1, 1, 3, 5, 5, 1, 3, 3, 5,
7, 6, 7]
```

Implementing a k-mer iterator

According to the method shown before for the enumeration of k-mers based on enhanced suffix arrays, a k-mer iterator has four internal fixed parameters that are the indexed string, the suffix array, the LCP array and the word length k. The variables of the internal state are the start and the end of the current LPC k-interval, identified by the private variables i and j respectively. Each time the `next` method is called on the iterator, the search for the successive interval starts. If two valid limits for the next interval are found, the iterator pauses the iteration and returns the k-mer corresponding to the found interval. If no more intervals are found, the iterator rises a `StopIteration` exception.

In addition, the iterator informs about the multiplicity of the current k-mer and the starting position of its occurrences along the string. The information is provided b the two methods `multiplicity` and `positions`, respectively. We recall, that the multiplicity is given by the size of the k-interval on the LCP array, and the set of occurring positions is the corresponding slice on the suffix array. Since the suffix array only ensures the lexicographical order of the suffix, the positions are not sorted.

Its definition follows the directives that are given in Sect. 7.11.

```
class ESAIterator:
    """
    An interator for extracting all the k-mers of a genomic
    sequence by means of the NESA data structure.
    """
    __s = None
    __k = 0
    __sa = None
    __lcp = None
    __i = 0
    __j = 0

    def __init__(self, s, k, sa = None, lcp = None):
        """
        If the suffix array (sa) and LCP array are not provided,
        then they will be computed by the constructor.
        """
        self.__s = s
        self.__k = k

        if sa == None:
            self.build_sa()
        else:
            self.__sa = sa

        if lcp == None:
            self.build_lcp()
        else:
            self.__lcp = lcp

    def build_sa(self):
```

```
        print("building suffix array...")
        suffixes = list()
        for i in range(len(self.__s)):
            suffixes.append( (self.__s[i:] + self.__s[:i] ,i) )
        self.__sa = list()
        for suff in sorted(suffixes):
            self.__sa.append(suff[1])
        print('done')
    def longest_prefix_length(s, i, j):
        l = 0
        while (i+l < len(s)) and (j+l < len(s)):
            if s[i+l] != s[j+l]:
                break
            l += 1
        return l

    def build_lcp(self):
        print('building lcp array...')
        self.__lcp = list()
        self.__lcp.append(0)
        for i in range(1,len(self.__sa)):
            self.__lcp.append(
             ESAIterator.longest_prefix_length(self.__s,
             self.__sa[i], self.__sa[i-1]) )
        print('done')

    def get_sa(self):
        return self.__sa

    def get_lcp(self):
        return self.__lcp

    def __iter__(self):
        return self

    def __next__(self):
        if self.__i < len(self.__s):
            self.__i = self.__j

            while (self.__i < len(self.__s)) and
                    (self.__sa[self.__i] >
                     len(self.__s) - self.__k - 1):
                self.__i += 1
            if self.__i == len(self.__s):
                raise StopIteration
            self.__j = self.__i+1
            while ( self.__j < len(self.__s) ) and
                (self.__lcp[self.__j] >= self.__k):
                self.__j += 1
            ret = self.__s[ self.__sa[self.__i] :
                    self.__sa[self.__i] + self.__k ]
            return ret
        else:
```

```
                raise StopIteration

        def multiplicity(self):
            return self.__j - self.__i

        def positions(self):
            return self.__sa[self.__i : self.__j]


it = ESAIterator('agctagctagctagtttagct', 3, sa=None, lcp=None)

for kmer in it:
    print(kmer, it.multiplicity(), it.positions())
```

```
building suffix array...
done
building lcp array...
done
agc 4 [17, 0, 4, 8]
agt 1 [12]
cta 3 [2, 6, 10]
gct 3 [1, 5, 9]
gtt 1 [13]
tag 1 [16]
tag 3 [3, 7, 11]
tta 1 [15]
ttt 1 [14]
```

In addition, the iterator builds by it self the suffix and the LCP arrays if they are not proved by the user. However, the construction of the arrays is an expensive step, thus, this is not a good practice if multiple iterators have to be launched on the same strings, even with different values of k. It is preferable to build the necessary data structure once and then, successively, to instantiate as iterators as needed on top of the same data structures.

```
s = 'agctagctagctagtttagct'

k = 2
it = ESAIterator(s, k, sa=None, lcp=None)

sa = it.get_sa()
lcp = it.get_lcp()

print(str(k)+"-mers")
for kmer in it:
    print(kmer, it.multiplicity(), it.positions())


k = 3
it = ESAIterator(s, k, sa, lcp)

print(str(k)+"-mers")
for kmer in it:
```

```
    print(kmer, it.multiplicity(), it.positions())
```

```
building suffix array...
done
building lcp array...
done
2-mers
ag 5 [17, 0, 4, 8, 12]
ct 3 [2, 6, 10]
gc 4 [18, 1, 5, 9]
gt 1 [13]
ta 1 [16]
ta 3 [3, 7, 11]
tt 2 [15, 14]
3-mers
agc 4 [17, 0, 4, 8]
agt 1 [12]
cta 3 [2, 6, 10]
gct 3 [1, 5, 9]
gtt 1 [13]
tag 1 [16]
tag 3 [3, 7, 11]
tta 1 [15]
ttt 1 [14]
```

There are many more efficient ways to build the suffix array and the LCP array. One of them is provided by the module `pysuffixarray`, which can be installed by the command `pip install pysuffixarray`. An example of how to build the data structures by means of this library is given in what follows.

```
from pysuffixarray.core import SuffixArray
sa_obj = SuffixArray(s)
sa_sa = sa_obj.suffix_array()
lcp_sa = sa_obj.longest_common_prefix()

print_sa_lcp(s,sa_sa,lcp_sa)
```

```
index   suffixes                SA  LCP
---------------------------------------
0                               21  0
1       agct                    17  0
2       agctagctagctagtttagct   0   4
3       agctagctagtttagct       4   10
4       agctagtttagct           8   6
5       agtttagct               12  2
6       ct                      19  0
7       ctagctagctagtttagct     2   2
8       ctagctagtttagct         6   8
9       ctagtttagct             10  4
10      gct                     18  0
11      gctagctagctagtttagct    1   3
12      gctagctagtttagct        5   9
13      gctagtttagct            9   5
```

```
14  gtttagct              13  1
15  t                     20  0
16  tagct                 16  1
17  tagctagctagtttagct     3  5
18  tagctagtttagct         7  7
19  tagtttagct            11  3
20  ttagct                15  1
21  tttagct               14  2
```

Because of technical details in the construction of the data structures by means of the library, the first position of the suffix array and the LCP array must be discarded if they are used in a k-mer iterator defined in the previous sections.

```python
from pysuffixarray.core import SuffixArray
sa_obj = SuffixArray(s)
sa_sa = sa_obj.suffix_array()[1:]
lcp_sa = sa_obj.longest_common_prefix()[1:]

k = 3
it = ESAIterator(s, k, sa_sa, lcp_sa)

print(str(k)+"–mers")
for kmer in it:
    print(kmer, it.multiplicity(), it.positions())
```

```
3–mers
agc 4 [17, 0, 4, 8]
agt 1 [12]
cta 3 [2, 6, 10]
gct 3 [1, 5, 9]
gtt 1 [13]
tag 4 [16, 3, 7, 11]
tta 1 [15]
ttt 1 [14]
```

Similarly to the iterator for the ESA data structure, an iterator based on the NESA data structure is implemented in the following source code. Differently from the previous version, this new iterator also implements the `reset` method that allows the iterator to reset to the initial state.

```python
class NESAIterator:
    __s = None
    __k = 0
    __sa = None
    __lcp = None
    __ns = None
    __i = 0
    __j = 0

    def __init__(self, s, k, sa = None, lcp = None, ns = None):
        self.__s = s
        self.__k = k

        if sa == None:
```

```
            self.build_sa()
        else:
            self.__sa = sa

        if lcp == None:
            self.build_lcp()
        else:
            self.__lcp = lcp

        if ns == None:
            self.build_ns()
        else:
            self.__ns = ns

    def get_k(self):
        return self.__k

    def reset(self):#←———
        self.__i = 0
        self.__j = 0

    def build_sa(self):
        print("building suffix array...")
        suffixes = list()
        for i in range(len(self.__s)):
            suffixes.append( (self.__s[i:] + self.__s[:i] ,i) )
        self.__sa = list()
        for suff in sorted(suffixes):
            self.__sa.append(suff[1])
        print('done')


    def longest_prefix_length(s, i, j):
        l = 0
        while (i+l < len(s)) and (j+l < len(s)):
            if s[i+l] != s[j+l]:
                break
            l += 1
        return l

    def build_lcp(self):
        print('building lcp array...')
        self.__lcp = list()
        self.__lcp.append(0)
        for i in range(1,len(self.__sa)):
            self.__lcp.append(
            NESAIterator.longest_prefix_length(
            self.__s, self.__sa[i], self.__sa[i-1]) )
        print('done')

    def build_ns(self):
        print('building ns array...')
        inv_sa = [0 for _ in range(len( self.__sa))]
```

```
        for i in range(len(self.__sa)):
            inv_sa[  self.__sa[i] ] = i

        self.__ns = [0 for _ in range(len( self.__sa))]
        lastn = len(self.__s)
        for i in range(len(self.__s)-1,-1,-1):
            if self.__s[i] == 'N':
                lastn = i
            self.__ns[ inv_sa[i] ] = lastn - i
        print('done')

    def get_sa(self):
        return self.__sa

    def get_lcp(self):
        return self.__lcp

    def get_ns(self):
        return self.__ns


    def __iter__(self):
        return self
    def __next__(self):
        if self.__i < len(self.__s):
            self.__i = self.__j

            while (self.__i < len(self.__s))
                    and  ( (self.__sa[self.__i] >
                        len(self.__s) - self.__k - 1)
                     or (self.__ns[self.__i] < self.__k) ):
                self.__i += 1
            if self.__i == len(self.__s):
                raise StopIteration
            self.__j = self.__i+1
            while ( self.__j < len(self.__s) )
                    and (self.__lcp[self.__j] >= self.__k)
                    and (self.__ns[self.__i] >= self.__k) :
                self.__j += 1
            ret = self.__s[ self.__sa[self.__i] :
                    self.__sa[self.__i] + self.__k ]
            #self.__i = self.__j #!!!!!!
            return ret
        else:
            raise StopIteration

    def multiplicity(self):
        return self.__j - self.__i

    def positions(self):
        return self.__sa[self.__i : self.__j]
```

```
it = NESAIterator('agctagctagNctagtttagctN', 3)

print('iterating '+str(it.get_k())+'-mers...')
for kmer in it:
    print(kmer, it.multiplicity(), it.positions())
```

```
building suffix array...
done
building lcp array...
done
building ns array...
done
iterating 3-mers...
agc 3 [18, 4, 0]
agt 1 [13]
cta 3 [6, 2, 11]
gct 3 [19, 5, 1]
gtt 1 [14]
tag 4 [7, 17, 3, 12]
tta 1 [16]
ttt 1 [15]
```

```
print('reiterating '+str(it.get_k())+'-mers...')
it.reset()
kmer = next(it)
print("first: ",kmer, it.multiplicity(), it.positions())

for kmer in it:
    print(kmer, it.multiplicity(), it.positions())
```

```
reiterating 3-mers...
first: agc 3 [18, 4, 0]
agt 1 [13]
cta 3 [6, 2, 11]
gct 3 [19, 5, 1]
gtt 1 [14]
tag 4 [7, 17, 3, 12]
tta 1 [16]
ttt 1 [15]
```

8.11 Recurrence Patterns

Given a word w, the recurrence distance distribution (RDD) informs about the distances at which w recurs in a reference string s. A recurrence distance is the distance between two consecutive occurrences of w. The RDD distribution reports how many times two consecutive occurrences recur at a given distance.

```
def RDD(s,w):
    """
```

```
        Extract the recurrence distance ditribution (RDD)
        of the word w in s.
        Given the starting positions of two occurences of w,
        p1 and p2, the reucrrence distance is calculated as
        p1 - p2 such that consecutive occurrences are
        at distance 1.
        ----
        Parameters:
            s (str) : the reference string
            w (str) : the searched substring
        ----
        Returns:
            dict[int,int] : a dictionary mapping recurrence
            distances to the number of times they occur
    """
    pos = sorted(get_positions(s,w))
    rdd = dict()
    for i in range(2,len(pos)):
        l = pos[i] - pos[i-1]
        rdd[l] = rdd.get(l,0) + 1
    return rdd

print(s+'\n')
print('RDD(a):', RDD(s,'a'))
print('RDD(t):',RDD(s,'t'))
print('RDD(c):',RDD(s,'c'))
print('RDD(g):',RDD(s,'g'))
```

```
agctaggaggatcgccagat

RDD(a): {2: 1, 3: 2, 6: 1}
RDD(t): {8: 1}
RDD(c): {1: 1, 2: 1}
RDD(g): {1: 2, 2: 1, 4: 2}
```

In real sequences, RDD distributions are usually defined on several values of recurrence distance. Thus, simple prints result in too much verbose output. The module `matlibplot` provide several functionalities for plotting charts in python. In this case, a bar plot is used to visualize the content of an RDD distribution.

If a recurrence distance between the minimum and the maximum retrieved values is missing, it is preferable to add such information by adding to the RDD a value equal to zero for the corresponding distances. In this way, the plot is more readable since the values on the x-axis are contiguous and no hidden gaps are produced.

In what follows, the RDD of the word *GC* is computed along the genome of the *M. genitalium* species.

```
w = 'GC'
rdd = RDD(mg37,w)

def plot_RDD(rdd, title):
    """
        Plot an RDD distribution adding missing recurring
```

```
        distances, between 1 and the original maximum distance,
        by adding a value of zero in correspondence of them.
    """
    # se a value equal to zero for the missing recurrence distances
    for d in range(0,max(rdd.keys())):
        rdd[d] = rdd.get(d,0) + 0

    # module can be imported by uring aliases to refer them
    import matplotlib.pyplot as plt

    # set the figure size
    plt.rcParams['figure.figsize'] = [20, 6]
    # assign height of bars
    bar_values = [v for k,v in sorted(rdd.items())]
    # plot with specific values on the x-axis that are
    # associated to the height
    plt.bar(sorted(rdd.keys()), bar_values, width=1.0)
    # set the label on the y-axis
    plt.ylabel('Number of pairs')
    # set the label on the x-axis
    plt.xlabel('Recurrence distance')
    # set a title for the chart
    plt.title(title)
    # plot the chart
    plt.show()

plot_RDD(rdd, 'Recurrence distance distribution of '+ w)
```

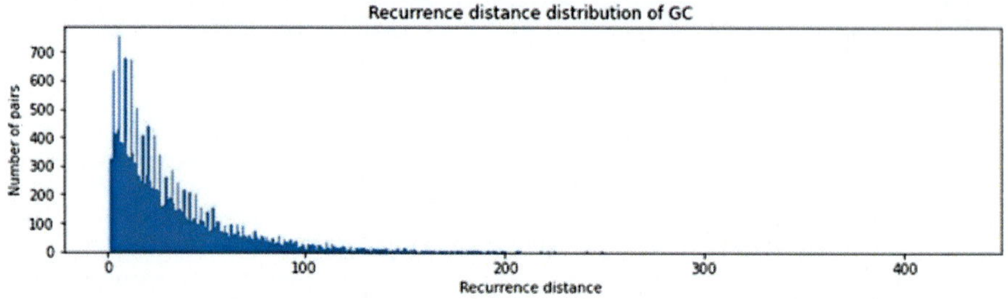

Average recurrence distance distribution

The average recurrence distribution (aRDD) reports an average value of recurrence regarding the RDDs of several words belonging to a dictionary. Usually, it is computed for the complete set of *k*-mers occurring in a string.

```
def aRDD(s,k):
    """
        Computer the average recurrence distance distribution of
        the complete set of k-mers occuring in s.
        ----
        Parameters:
```

```
            s (str) : the string from which extract the RDDs
            k (int) : the word length of the k-mers for which
                        extract the RDD

        ----
        Returns:
            dict[int,float] : a dictionary mapping recurrence
                        distances to the average number of
                        times they occur
    """
    ardd = dict()
    kmers = get_kmers(s,k)
    for kmer in kmers:
        rdd = RDD(s,kmer)
        for distance,value in rdd.items():
            ardd[distance] = ardd.get(distance,0) + value
    for d,v in ardd.items():
        ardd[d] = ardd[d] / len(kmers)
    return ardd
```

The following code shows the aRDD between the recurrence distances 1 and 60.

```
nardd = dict()
for d,v in ardd.items():
    if d <=  60:
        nardd[d] = v
ardd = nardd

plot_RDD(ardd, 'Average recurrence distance distribution for k '+
 str(k))
```

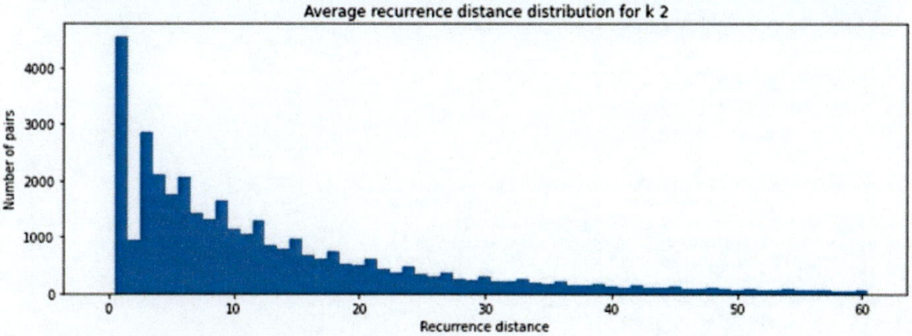

Reading GFF3 files

The GFF3 (general feature format, version 3) file format is sued to specify a list of annotations along a genome. An annotation is a contiguous region of the genomic sequences which show some particular properties. Common annotations are the genes. Each gene is identified by the starting and the ending coordinate, the strand and some supplementary non-physical descriptions.

GFF3 files are comma-separated values (CSV) files, which means that they are textual files in which a sort of column content is represented. Columns are separated with a specific character, usually a comma o a tabulation character. Additional comment lines are identified with an initial # character.
GFF3 have at least 6 tab-separated columns that are:

- 0 : annotation identifier
- 1 : source (database, organization, etc...)
- 2 : feature type (region, gene, exon, tRNA, etc...)
- 3 : start coordinate
- 4 : end coordinate
- 5 : score (gives an indication regarding the confidence of the sources)
- 6 : strand (+ or –)
- 7 : phase (phase o- CDS features)
- 8 : additional attributes

Independently from the strand, coordinates are given as absolute values from the first nucleotide on the (+) 5'–3' strand.

The following code reads the GFF3 annotation file of the mycoplasma genitalium genome in order to extract the parts of it that are covered by gene annotations. Since genes may reside on both strands and they also may overlap, thus the sequence coverage is intended as the number of positions in the 5'–3' strand that are covered by at least one gene independently of the strand.

We also recall that GFF3 file usually only reports annotation coordinates, thus the actual sequence must be retrieved from a corresponding FASTA file.

```
ifile = 'mycoplasma_genitalium_G37.fna'
mg37 = ''
for line in open(ifile, 'r'):
    if line.strip()[0] != '>':
        mg37 += line.strip()

coverage = [0 for i in range(len(mg37))]

ifile = 'mycoplasma_genitalium_G37.gff3'
for line in open(ifile, 'r'):
    if line[0] != "#":
        cc = line.split('\t')
        if len(cc) >= 6:
            if (cc[2] == 'gene'):# and (cc[6] == '+'): # we
calculate the coverage of both strands as a single strand
                start = int(cc[3])
                end = int(cc[4])
                for i in range(start, end):
                    coverage[i] += 1

print('sequence coverage',
 (len(coverage) - coverage.count(0)) / len(coverage))

# sequence of non-coding portion of the genome
```

```
ncseq = ''.join(
 [ mg37[i] for i in range(len(mg37)) if coverage[i] == 0 ])

# sequence of coding portion of the genome
cseq = ''.join(
 [ mg37[i] for i in range(len(mg37)) if coverage[i] > 0 ])

print('total length', len(mg37),', non-coding length',len(ncseq),
 ', protein-coding length', len(cseq))
```

```
sequence coverage 0.9502547941993807
total length 580076, non-coding length 28856, protein-coding
length 551220
```

The sequence coverage indicates that 95% of the mycolpasma genitalium is covered by coding genes. This has crucial implications for the composition of the genome. In fact, it means that it is almost composed of nucleotides that code for codons. Of course, the k-mer content is expected to be slightly different from the non-coding portions.

The following is the multiplicity distribution of all the theoretical 2-mers computed on the non-coding part.

```
def list_words_2(prefix, k, nuc_alphabet, words):
    if len(prefix) == k:
        words.add(prefix)
    else:
        for a in nuc_alphabet:
            list_words_2(prefix + a, k, nuc_alphabet, words)

k = 2
kmers = set()
list_words_2('',k,['A','T','C','G'], kmers)
kmers = sorted(kmers)

ncseq_wmd = dict()
for kmer in kmers:
    ncseq_wmd[kmer] = count_occurrences(ncseq, kmer)

import matplotlib.pyplot as plt
bar_values = [v for k,v in sorted(ncseq_wmd.items())]
plt.rcParams['figure.figsize'] = [20, 6]
plt.bar(kmers, bar_values)
plt.xticks(kmers, kmers)
plt.ylabel('multiplicity')
plt.xlabel('words')
plt.title('Word multiplicity distribution')
plt.show()
```

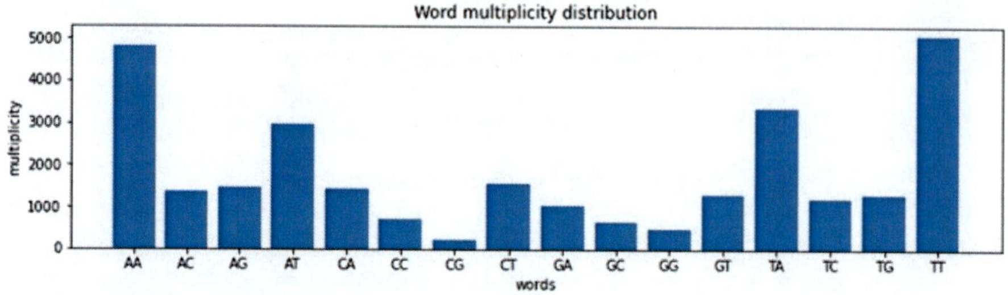

The following is the multiplicity distribution of all the theoretical 2-mers computed on the coding part.

```
cseq_wmd = dict()
for kmer in kmers:
    cseq_wmd[kmer] = count_occurrences(cseq,kmer)

import matplotlib.pyplot as plt
bar_values = [v for k,v in sorted(cseq_wmd.items())]
plt.rcParams['figure.figsize'] = [20, 6]
plt.bar(kmers, bar_values)
plt.xticks(kmers, kmers)
plt.ylabel('multiplicity')
plt.xlabel('words')
plt.title('Word multiplicity distribution')
plt.show()
```

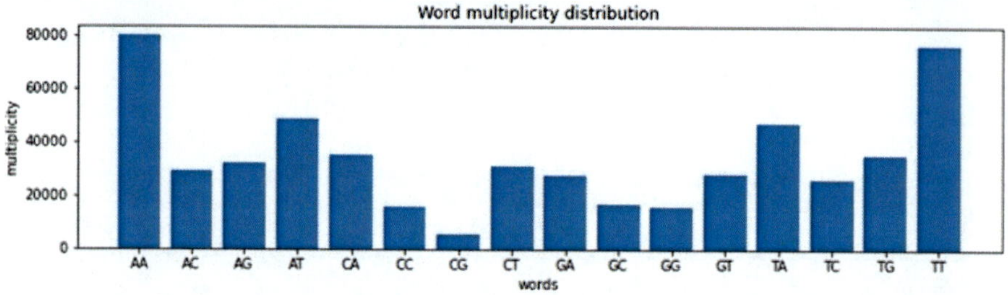

There are differences between the two charts, however, there is not a strong signal that indicates the presence of the coding language. On the contrary, there is strong evidence of the codon language via the analysis of the RDD composition. In the next sections, a crucial aspect of RDDs extracted from coding regions is distinctive of such parts w.r.t. to non-coding regions.

Differences between coding and non-coding parts

The following chart is the average RDD calculated for the overall mycoplasma genitalium genome, as has been shown before. The distribution seems to follow an exponential distribution, as expected, but some distinctive peaks on recurrence distance multiple of 3 appear. Such peaks seem to follow a parallel exponential distribution.

```
k = 2
ardd = aRDD(mg37,k)

nardd = dict()
for d,v in ardd.items():
    if d <  100:
        nardd[d] = v
ardd = nardd

plot_RDD(ardd, 'Average recurrence distance distribution for k '
 + str(k))
```

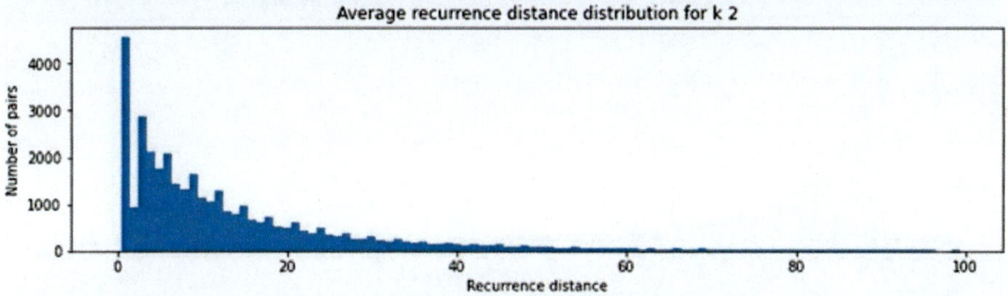

The following chart is the average RDD calculated over only the non-coding part of the genome. The parallel 3-peak distribution disappears.

```
k = 2
ardd = aRDD(ncseq,k)

nardd = dict()
for d,v in ardd.items():
    if d < 100:
        nardd[d] = v
ardd = nardd

plot_RDD(ardd, 'Average recurrence distance distribution on
non-coding parts for k '+ str(k))
```

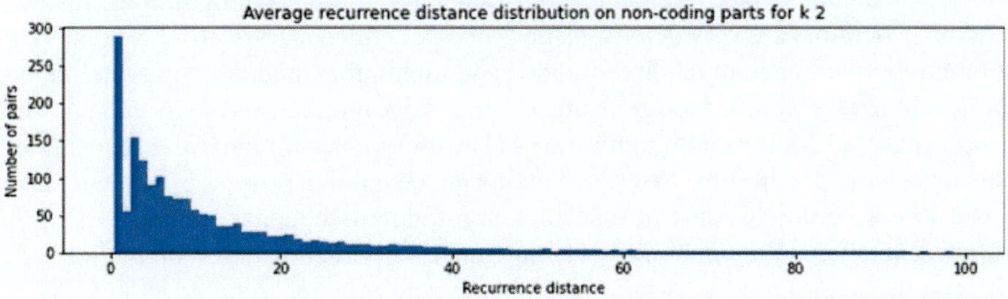

When the average RDD is computed on only the coding part of the genome, the parallel 3-peak distribution appears back. The phenomenon is given by the 3-periodicity of the codon language that is reflected in the 3-peak distribution of the RDDs of almost all the 2-mers. The implications on the overall sequence lead to the fact that the genome is almost covered by coding sequences

```
k = 2
ardd = aRDD(cseq,k)

nardd = dict()
for d,v in ardd.items():
    if d < 100:
        nardd[d] = v
ardd = nardd

plot_RDD(ardd, 'Average recurrence distance distribution on
coding parts for k '+ str(k))
```

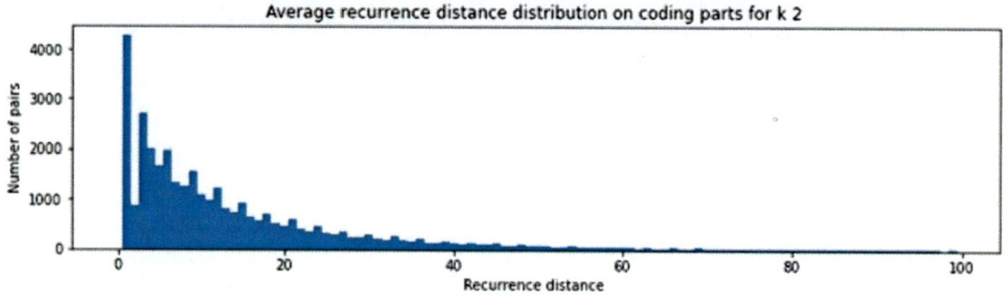

8.12 Generation of Random Genomes

This section shows how to produce random strings in `python3`. The basic approach uses the pseudo-random number generator provided by the built-in library of python called `random`. The library implements several methods for generating random integers which can also be exploited for surrogate operations such as shuffling a sequence of elements or picking up a random element from k elements from of a given alphabet of allowed elements symbols, by generating a casual sequence. However, one of the most important functions of this library does not strictly concern the output of the random generator, but rather its internal configuration. In fact, random generators are very deterministic processes which start from a given internal configuration to generate the first pseudo-random number, and then by switching to the next internal configuration generate the successive number, and so forth. Thus, the parameters of the initial configuration are of key importance for the entire sequence of generated numbers. In some cases, depending on the internal process of the generator, a bad initial configuration can lead to a non-random sequence of numbers. One famous example is the linear congruential generator which is based on the formula $X_{n+1} = (aX_n + c) \bmod m$, where X_{n+1} is the next number to be generated, X_n is the number generated at the previous step and a, c and m are the parameters of the generator. It can be shown that for $X_0 = 1, a = 3, c = 1, m = 5$ the generator produces a sequence with period 4. Modern generators, such as the Mersenne Twister generator, are much more robust and safe, but still, the initial configuration plays an important role in the quality of randomness level of their generations. Random sequences as well as random modifications of a biological sequence are used to evaluate the divergence from the randomness of the real biological sequences. Often, empirical p-values are calculated by randomly shuffling sequences or by generating random instances. If different sequences are generated by different executions of the same script, then very likely different results will be obtained. thus, it is of primary importance to ensure the reproducibility of the results, which means that multiple executions of the same script must return the same result. If the script contains the generation of random sequences, or in general random elements, these elements must be the same in every execution of the script. In this context, reproducibility is ensured by setting the initial configuration of the random generator via the function `random.seed`, which takes as input a number, called seed, to set up the initial configuration of the generator.

A first approach to generate random sequences of nucleotides, thus genomes, is by emulating a Bernoullian process in which the four nucleobases have the same probability to be extracted. This is obtained by taking into account that the basis random generator works by generating numbers in the range 0 to 2^b, where b is the number of bits of the architecture that is running the code, with uniform distribution. It is trivial to convert numbers generated in the interval $[0, 2^b]$ to numbers in the interval $[x, y]$, where x and y are user-specified limits of the interval. Thus, one can ask to generate numbers in [0, 3] and then convert them into the 4 nucleobases by means of a bijective function between [0, 3] and $\{A, C, G, T\}$, as it is shown in the following example.

```
import random
random.seed(123456)

map = {0:‘A’, 1:‘C’, 2:‘G’, 3:‘T’}

s = ” #an empty string to be filled
for i in range(100):
    s += map[ random.int(0,3) ]

print(s)
```

The script uses `123456` as seed and always generates the same sequence of 100 nucleotides. It has to be pointed out that such a sequence is constant between multiple executions of the scripts in the same environment. This means that different versions of pyhton, as well as different versions of the `random` library, may generate different results. For this reason, it is crucial to make the script publicly available when publishing the results of a study. and it is of the same importance to report any detail regarding the environment, especially the version of each library that was used.

An alternative way to generate the same sequence is given by using the function `random.choice` which randomly picks up an element of the input list.

```
random.seed(123456)

alphabet = [‘A’,‘C’,‘G’,‘T’]

s = ”
```

```
for i in range(100):
    s += random.choice(alphabet)

print(s)
```

Since the two scripts generate the same sequence, it is intuitive to think that the internal functioning of `random.choice` is very close to the first example, but two important aspects must be considered. Firstly, both scripts set the same seed for the random generator. Secondly, the order in which the symbols are declared in the list `alphabet` follows the mapping given by the dictionary `map` of the first script.

In real cases, the Bernoullian uniform process can not be the most suitable model for generating random strings. In particular, if the frequencies of the nucleobases are taken into account. In real genomes, the nucleobases do have not the same frequency, and one may want to generate a random string which at least respects such frequencies. A possible approach is to implement a Bernoullian process in which a different number of balls are used to fill the urn. Thus, if the probabilities of nucleobases are expressed in cents, then an alphabet/list of 100 symbols can be passed to the `random.choice`. Such a list contains a number of A elements that is proportional to the probability of A, as well as for the other 3 symbols.

```
import math
import random

random.seed(123456)

resolution = 100

probabilities = {'A':0.2, 'C':0.3,'G':0.3,'T':0.2}

alphabet = []
for k,v in probabilities.items():
    alphabet += k*math.ceil(resolution * v)

s = ''

for i in range(100):
    s += random.choice(alphabet)
```

The example can be used to generate sequences of any length by keeping the same resolution for the probabilities.

It has to be noticed that such an approach does not ensure that the real and the random strings have the same multiplicity distribution. In fact, given two k-mers, for example $\alpha = a_1a_2$ and $\beta = b_1b_2$, with $a_1, a_2, b_1, b_2 \in \{A, C, G, T\}$, their consecutive extraction produces the substring $\alpha\beta = a_1a_2b_1b_2$ and thus the unwanted a_2b_1 2-mer. Thus, it can be better to model the process with a Markov chain such that it assigns a probability of extracting a symbol from the alphabet by looking at the last generated nucleotide. In any case, none of the two approaches can guarantee the exact correspondence between the observed frequency of the k-mers in the real genome and the random string. Such approaches are then helpful in studying the random composition of k-mers starting from frequencies of $k - 1$-mers. An alternative way, which preserves k-multiciplicities, is to build a de Bruijn graph of order $k - 1$ as a multigraph in which the number of edges between two vertices equals the multiplicity of the k-mer formed by the junction of the two $(k - 1)$-mers represented by the vertices. Thus, the solution is obtained by finding all the Eulerian paths which start from the first k-mer of the real genome, and then randomly selecting one of them. This approach could be employed in the recurrences of words rather than the overall multiplicity distribution.

References

1. Aardenne, T., de Bruijn, N.G.: Circuits and trees in oriented linear graphs. Simon Stevin **28**, 203–217 (1951)
2. Bonnici, G.R., Manca, P.: A dictionary-based method for pan-genome content discovery. BMC Bioinform. **19**(Suppl 15), 437 (2018)
3. Chaitin, G.J.: On the simplicity and speed of programs for computing infinite sets of natural numbers. JACM **16**(3), 407–422 (1969)
4. Chaitin, G.J.: Algorithmic Information Theory. Cambridge University Press (1987)
5. Church, A.: On the concept of a random sequence. Bull. Am. Math. Soc. **46**, 130–135 (1940)
6. Devroye, L.: Non-Uniform Random Variate Generation. Springer, Berlin (1986)
7. Downey, R.G., Hirschfeldt, D.R.: Algorithmic Randomness and Complexity. Springer, Berlin (2010)
8. Holland, J., Mallot, H.: Emergence: from chaos to order. Nature **395**, 342–342 (1998)
9. Knuth, D.: The Art of Computer Programming Volume 1 (Fundamental algorithms) and Volume 2 (Semi-numerical algorithms). Addison-Wesley (1997)
10. Kolmogorov, A.N.: Logical basis for information theory and probability theory. IEEE Trans. Inf. Theory **14**, 662–664 (1968)
11. Kolmogorov, A.N.: On tables of random numbers. Theor. Comput. Sci. **207**(2), 387–395 (1998)
12. Kullback, S., Leibler, R.A.: On information and sufficiency. Ann. Math. Stat. 79–86 (1951)
13. Landauer, R.: Dissipation and noise immunity in computation and communication. Nature **335**, 779–784 (1988)
14. Li M., Vitanyi, P.: An Introduction to Kolmogorov Complexity and Its Applications. Springer, Berlin (2008)
15. Manca, V.: Logica Matematica. Bollati-Boringhieri (2005)
16. Manca, V.: Linguaggi e Calcoli. Principi matematici del "coding". Bollati Boringhieri (2019)
17. Manca, V.: Formule e Metafore. Immaginazione e Concettualizzaaione scientifica, Edizioni Nuova Cultura (2021)
18. Manca, V., Santagata, M.: Un meraviglioso accidente. La nascita della vita, Mondadori (2018)
19. Manca, V., Bianco, L.: Biological networks in metabolic P systems. BioSystems **91**, 489–498 (2008)
20. May, R.M.: Simple mathematical models with very complicated dynamics. Nature **261**, 459–67 (1976)
21. McEliece, R.J.: The Theory of Information and Coding. Cambridge University Press (2003)
22. Pincus, S.M.: Approximate entropy as a measure of system complexity. P. Nat. Acad. Sci. **88**, 2297–2301 (1991)

V. Manca and V. Bonnici, *Infogenomics*, Emergence, Complexity and Computation 48,
https://doi.org/10.1007/978-3-031-44501-9

23. Rosenblueth, A., Wiener, N., Bigelow, J.: Behavior, purpose and teleology. Philos. Sci. **10**(1), 18–24 (1943)
24. Solomonov, R.: An exact method for the computation of the connectivity of random nets. Bull. Math. Biophys. **14**, 153 (1952)
25. Soto, J.: Statistical Testing of Random Number Generators (1999). http://csrc.nist.gov/rng/

Index

V. Manca and V. Bonnici, *Infogenomics*, Emergence, Complexity and Computation 48,
https://doi.org/10.1007/978-3-031-44501-9